# Symmes's Theory of Spheres

Demonstrating that the Earth is hollow, habitable within, and widely open about the poles

James McBride and John Cleves Symmes

**Alpha Editions**

This edition published in 2024

ISBN : 9789366381596

Design and Setting By
**Alpha Editions**
www.alphaedis.com
Email - info@alphaedis.com

As per information held with us this book is in Public Domain.
This book is a reproduction of an important historical work. Alpha Editions uses the best technology to reproduce historical work in the same manner it was first published to preserve its original nature. Any marks or number seen are left intentionally to preserve its true form.

# Contents

DISTRICT OF OHIO, to wit. ................................................- 1 -

ADVERTISEMENT. ...............................................................- 2 -

To the Public. ..........................................................................- 3 -

Preface. ...................................................................................- 4 -

Apology TO CAPTAIN SYMMES. ......................................- 6 -

CHAPTER I. ..........................................................................- 7 -

CHAPTER II. .......................................................................- 13 -

CHAPTER III. ......................................................................- 19 -

CHAPTER IV. ......................................................................- 27 -

CHAPTER V. .......................................................................- 32 -

CHAPTER VI. ......................................................................- 42 -

CHAPTER VII. .....................................................................- 47 -

CHAPTER VIII. ...................................................................- 55 -

CHAPTER IX. ......................................................................- 62 -

CHAPTER X. .......................................................................- 66 -

FOOTNOTES ......................................................................- 70 -

# DISTRICT OF OHIO, TO WIT.

BE IT REMEMBERED, that on the fourth day of April, in the year of our Lord one thousand eight hundred and twenty six and in the fiftieth year of the American independence, MESSRS. MORGAN, LODGE AND FISHER, of said District, hath deposited in this office, the title of a book, the right whereof they claim as proprietors, in the words and figures following, to wit:

"Symmes's theory of concentric spheres; demonstrating that the earth is hollow, habitable within, and widely open about the poles: by a citizen of the United States. 'There are more things in Heaven and Earth Horatio, than are dreamt of in your philosophy' Shakespeare, 'If this man be erroneous who appears to be so sanguine and persevering in his opinions, what withholds us but our sloth, our self-will, and distrust in the right cause, that we do not give him gentle meetings and a gentle dismission; that we debate not and examine the matter thoroughly, with liberal and frequent audience: if not for his sake, yet for our own; seeing that no man who has tasted learning but will confess the many ways of profiting by those, who, not content with stale receipts, are able to manage and set forth new positions to the world. And were they but as the dust and cinders of our feet, so long as in that notion, they may yet serve to polish and brighten the armory of truth: even for that respect, they are not utterly to be cast away.' Milton."

In conformity to the act of Congress of the United States, entitled "An act for the encouragement of learning by securing the copies of Maps, Charts and Books to the proprietors of such copies during the times therein mentioned;" and also of the act entitled "An act supplementary to an act entitled an act for the encouragement of learning by securing the copies of Maps, Charts and Books, to the authors and proprietors of such copies, during the times therein mentioned, and extending the benefits thereof to the arts of designing, engraving and etching historical and other prints."

Attest, WILLIAM KEY BOND, CLERK.

## ADVERTISEMENT.

The writer of the following work is said to be a resident of the Miami country. After reading Captain Symmes's numbers, and hearing some of his lectures, he wrote the work, it seems, in the first place without the idea of publication; but afterwards corrected and enlarged it, and left it with a friend of Captain Symmes for publication, sometime in the autumn of the year 1824. The nett profits were then, as now, to be paid to Captain Symmes, towards enabling him to promote and establish his principles: but owing to the absence of the author, and other circumstances, it has remained unpublished till now.

The author has chosen to present the work anonymously; and has obtained the promise of Captain Symmes to forbear criticising it in manuscript,—reserving any remarks or corrections, he may wish to make, for future publication. Some *errors of the press* will doubtless be discovered; as (in the absence of both Compiler and Theorist) there was no *proof-reader* at hand, sufficiently versed in the New Theory, at all times, to detect them.

THE PUBLISHERS.

*Cincinnati, April, 1826.*

## To the Public.

The following little treatise, was written in the autumn of the year eighteen hundred and twenty-four; when from the urgency of my common avocation, and from a desire to remain *incognito*, the manuscript was placed in the hands of a friend of Captain Symmes for publication. As it was not my intention to seek a publisher, or make advances to facilitate its progress, I left the country for a considerable length of time, without paying any further attention to the subject. Various difficulties intervening, delayed the publication, until *subsequent events*, have destroyed my chief inducement; which was, that these speculations, compiled from a cursory examination of facts, should go forth as a harbinger, merely, and not "*follow in the wake*," of public investigation.

THE AUTHOR.

*March, 1826.*

# Preface.

The author of the following pages does not write because he is a learned man; he is conscious of the reverse; and that his merits give him no claim to that appellation; neither does he make this attempt because he is well acquainted with either the new, or the old theories of the earth; but, from having observed that the Theory of Concentric Spheres has been before the world for six or seven years, without attracting the attention of the scientific, except in a very few instances;—few besides the author himself having come forward to advocate its correctness. The newspaper scribblers, who have noticed the theory at all, have almost uniformly appeared to consider it as a fit subject on which to indulge their wit, the sallies of which, clothed in all the humour and satire their fancies could suggest, have in some degree had a tendency to throw around it an air of levity very unfavourable to serious investigation. But to deal in sarcasm is not always reasoning; and the truth is not to be ascertained by indulging in ridicule.

Considerations of this nature, first induced the author to devote a short time to the task of investigating a subject, to which he had paid but little attention, and to give the several papers, published by Captain Symmes, a cursory examination; in the course of which, he noted such of Symmes's principles and proofs as attracted his attention, as they occurred; and has since presumed to arrange them in such order as his own fancy suggested; supposing that, as they had struck forcibly on his mind, they might perhaps attract the attention of some other person, whose habits of thinking may be similar to his own. He has in a few instances inserted, in addition to those which he has seen advanced by Captain Symmes, such reasons and proofs in support of the theory as occurred to him at the time. However, he has no claim to originality; as he has made a liberal use of the publications of Captain Symmes, as well as the remarks made on them by others, which came in his way.

The reader will not look for a complete analysis of the theory in this short treatise; it is not intended as such by the author, his object being merely to attract the attention of the learned, who are in the habit of indulging in more abstruse researches into the operation and effect of natural causes; and should it be found to merit the attention of such, it is hoped their enquiries may be so directed as to accelerate the march of scientific improvement, enlarge the field of philosophic speculation, and open to the world new objects of ambition and enterprise.

Should he therefore be fortunate enough to make any observations, or indulge in any reflections, in the course of the following chapters, that may merit the attention of the reader, he hopes they may in some degree atone for the many defects which will doubtless be discovered; with a sincere wish, that gentlemen of literature and science, who have made deeper researches than he pretends to, will have the goodness to correct them.

The author does not write for Fame: as anonymous compilers (and it is the author's wish to be considered in no other light) can never expect their true names to be inscribed on her records: neither do pecuniary considerations influence him, as he expects to reap no profit from the publication.

Should it attract public curiosity to such a degree, as to induce the sale of more copies than will be sufficient to meet the expense of printing, it is the author's desire, and he does hereby direct, and fully authorize the publishers, to pay over the nett profits to Captain Symmes, for the purpose of enabling him further to prosecute his studies; and to aid him in the accomplishment of his designs.

Whether Captain Symmes has hit upon an important truth in the economy of nature, as respects the organization of matter, it is not for the author to determine; to the more scientific we must look for a solution of the problem; to them it is submitted. The following pages are presented with no other intention, than as a hint to elicit the attention of others, who are qualified to investigate, and improve the subject. Should they, on examination, consider the matter worthy of their investigation, it will doubtless receive the attention which its importance so greatly demands. If it be erroneous, it is hoped they will detect, and expose its fallacy to the world; giving at the same time rational and satisfactory explanations of the many facts, and appearances which Captain Symmes adduces as proofs of his positions.

*August, A. D. 1824.*

# Apology
# TO
# CAPTAIN SYMMES.

SIR—

To you I would apologize for the liberties I have taken with your Theory, and your publications in relation to it, which have made their appearance in the newspapers of the day. When I commenced this compilation, in support of your doctrine of Concentric Spheres, I had no view to its publication. I had collected all the papers on the subject, upon which I could lay my hands, with the intention of investigating the Theory for my own satisfaction: but the scattered and irregular order in which I found them, and in which they must necessarily appear in detached Newspaper essays, published at different and distant times, induced me to attempt a methodical arrangement, for the purpose of facilitating my own enquiries. When I had completed this, the same reasons, added to the consideration, that you have not only invited, but solicited the investigation of your theory, declaring it "as free as air," to every person, to make such use of it as he may think proper, influenced me to conclude on publishing the result of my investigations. Having come to this determination, I have added a Preface, an Introductory chapter, and a few things in conclusion, to make it look more like a Book.

As I have not seen all your publications in the newspapers, if I have not fully understood, or if I have misrepresented your theory in any particular, I assure you it has been done unintentionally—it has arisen entirely from my want of adequate information; and I hope you will, in the spirit of candour and good nature, pardon and correct any errors into which I may have fallen. Had an opportunity offered, and could I have done it with propriety, I should certainly have submitted the manuscript to your revision, previous to its publication. However, as this sketch is only intended to elicit further investigation, and can only live until a formal and systematic treatise shall appear from your pen, I hope you will permit it to pass as the Pioneer to a more complete demonstration of your Theory of Concentric Spheres.

    I AM SIR,

        *One of the believers in that Theory,—*

THE AUTHOR.

1824.

# CHAPTER I.

*Containing an introductory glance at some of the different Theories and Opinions which have been advanced respecting the formation of the Earth, and the reception which those Theories met with from the world when first promulgated.*

It often happens, that those who have been early taught to believe a certain set of principles and doctrines as true, whether in philosophy, religion, or politics, adhere to them with the utmost pertinacity during the remainder of their lives. Any new theory, or principle, is resisted with peculiar energy; and, however inconsistent or untrue their favorite systems may be, they are disposed to make principles and facts bend to them; and would sooner call in question the general and immutable laws of nature, than the correctness of their own opinions. Perhaps this pertinacious adherence to prevalent and received opinions has retarded the progress of philosophic improvement more than the want of bold, original, and enquiring genius.

In former times those who cultivated science, or rather those who were called learned, generally based their philosophy on the doctrines of Aristotle; which, as they had been taught to reverence them from their infancy, had become almost interwoven with their constitutions. Hence, though time has unfolded to us their errors, during several centuries, suspicion never hinted their fallibility. The doctrine of the revolutions of the earth, and other planets; of gravitation, magnetism, and other properties now known to belong to matter; have each in their turn met with a strong opposition from the most learned men living at the time of their discovery. But, notwithstanding this opposition, in all ages, a few bold, enquiring minds have had the firmness to dissent from the established doctrines of the schoolmen, and to lay the foundation of new systems, the correctness of which subsequent improvements in science have more or less demonstrated to the world.

Although nearly six thousand years have elapsed since man has been placed upon the earth, he yet knows but little of its formation. Notwithstanding all our enterprise, all our boasted acquirements, and discoveries, its true form yet remains uncertain; and although admitted that it is not quite eight thousand miles in diameter, we still have never explored its extent. A space of nearly forty degrees of latitude remains as little known to us, as if it were a part of the surface of Saturn, or an orb revolving round a star of the eighth magnitude. We know nothing of the inhabitants of those regions, or what kind of animate beings exist in them.

It was a prevailing opinion among the ancients, the correctness of which they for ages never called in question, that the temperate zones of our globe were alone habitable.—The torrid zone they imagined was composed of nothing but sandy deserts, scorched up by the vertical and insupportable beams of a burning sun. The frigid zones, they believed were begirt with eternal snows, and "thick ribbed ice," which rendered them inaccessible to man, and incapable of supporting animal or vegetable life. Hence none ventured to approach them.

Subsequent discoveries have, however, taught us the errors of the ancients. We now know that the torrid zone teems with organic life; and possesses, in many parts, a population more dense than the temperate, and is equally well adapted to its support: nay, we even find the temperature of that region to be such that it contains mountains capped with perpetual snows, which the beams of a July sun do not dissolve. It has also been ascertained that the frigid zones are partially inhabited: but it seems that a certain timid dread, perhaps in part attributable to the prejudices imbibed from our ancestors, has prevented our exploring the extent of those regions. However, as far as civilized man has yet ventured to penetrate towards the poles, we find that plants grow, flowers bloom, and human beings make a permanent residence; nay, even the untutored savages who reside there tell us that other human beings reside yet further to the north; and animals are known to migrate in that direction. Reasoning then from analogy, and from what we know, we have no ground to conclude that such a vast extent of surface has been created by an all-wise Providence for no other purpose, than to be eternally clothed with mountains of ice. Such a conclusion comports not with the general economy *we do know* to exist throughout his works.


We are constrained to acknowledge, notwithstanding our improvements in science, that, comparatively, we know but little of the economy of nature. Within a few years past, almost an entire revolution has taken place in the world respecting the philosophy of light and heat—a change which affects the theory both of their nature, and of their causes:—They are now believed to be two distinct things, and that the sun communicates neither, but merely gives activity, in some manner not yet known, to the principles, or matter, of light and heat with which our elements abound. If this be the case, as I believe is now admitted by the learned world, we cannot undertake to say, that the intensity or the absence of either, is necessarily dependant alone on the altitude of the sun, under any particular latitude; or on our nearness to, or remoteness from, the centre of the system:—For aught we know, both may be connected with arrangements that require but few of the sun's rays to make them answer the purposes of organic life. For aught we can tell, the planet Georgium Sidus, which rolls eighteen hundred millions of miles distant from the orb of day, may, nevertheless, be favoured with as brilliant light, and as genial warmth as our little globe; and for aught we know the interior of this planet, in the concavity of the spheres, under the equator, may enjoy the same light and heat that fructify and bless the equatorial climes on the convex surface.

During a period of several thousand years the ancients were of opinion that the earth was a perfect plane, at rest, and supported below by an unknown something; that it was bounded on all sides by an impassable barrier, and covered with the blue canopy of heaven, in which the sun, moon, and stars performed their diurnal revolutions for the sole use and service of a few frail mortals. They believed that the sun, every morning rose out of the Eastern sea; and in the evening plunged into the Western ocean; that the stars were lighted up in the evening by some kind deity, and extinguished before the appearance of the sun. For ages none doubted the correctness of such a theory. At length, however, from an attentive examination of the regular appearances and revolutions of the heavenly bodies, some of the Babylonians adopted the opinion that the earth was spherical; revolving at regular periods round the sun, as the centre of the universe. In this they were followed by Pythagoras and others. But those efforts of genius, for the most part, met no other reward than the execrations of the exasperated multitude. Such innovations were deemed an impious crime against the gods, and could only be atoned for by the sacrifice of their lives. In those times the people of every nation, like the untutored Indian of our North Western wilderness at this day, considered their own country to be situated in the centre of the world, and they, the most favoured people. Even in later times, when the system of the Babylonians, and that of Pythagoras, were revived by Copernicus; and, when new discoveries respecting the form and revolutions of the earth, and other parts of the universe, were made by Galileo, not more than two hundred years since, we find an ignorant and bigoted world alarmed at such opinions. We find Galileo, that incomparable philosopher, cited before the court of Inquisition, accused of heresy, and thrown into prison. The charge of heresy against him was supported by alleging that he maintained the two following positions, viz.

1. "That the sun is the centre of the world, and immoveable by a local motion;" and


2. "That the earth is not the centre of the world, nor immoveable, but that it moves with a diurnal motion."

These positions he was not permitted to maintain or defend, but was ordered to renounce them; and was prohibited from vindicating them either in conversation or writing. However strange and impious these doctrines appeared at that time, subsequent ages have confirmed their correctness.

When Columbus advanced the theory of a western continent, he was ridiculed, persecuted, and contemned, by nearly all the literati of Europe. It was an idea which had never before entered their minds. But, notwithstanding all their opposition and ridicule, the correctness of his "visionary theory," as they were pleased to call it, was demonstrated by the actual discovery of this vast continent, which is now sustaining millions of the very happiest of the human race.

Many of the important discoveries of the immortal Newton, at the time they were first promulgated to the world, were denounced as the splendid visions of a madman; but, subsequent ages have done him justice.

Much as we may feel ourselves elated on account of the new lights which have since been shed upon us, by the further progress and developement of science; yet, when I reflect on the unkind treatment which Captain Symmes and his new theory have received in our own day, I cannot help fearing that we are still, in some degree, under the influence of the same feelings and prejudices which brought the earlier philosophers to the torture, and the prison. This theory differs much less from the one now commonly received, than the doctrines of those philosophers differed from the prejudices of the multitude, in an age when every one believed the earth to be as flat as a table; and, consequently, it is but a small innovation in comparison to what the theory of Pythagoras and Copernicus must have appeared to be in their day; yet Captain Symmes has been constantly, and almost every where, represented as a visionary and dangerous innovator, and his alleged discovery ridiculed as the silly dream of a deranged imagination.

But let us not turn our backs and give a deaf ear to him, or to the discoveries of any other man, merely because they are new, and in contravention of our previously received impressions. True it is, novelty is frequently dangerous and hurtful: but on the other hand, it is often necessary and useful. Without it we should still remain destitute of many of the greatest advantages we enjoy. Without the advancement of new principles, and speculative ideas, neither ourselves, nor any other people, could ever have emerged from a state of savage barbarity. Without it, what purpose could our reason serve, which, under proper regulations, and by a gradual progress, is capable of contributing so largely to the general good of society?

Were it my opinion that Symmes's Theory is one of the wildest and most ridiculous that ever entered into the brain of man, I would not refuse to hear him; nor by malevolent or satirical disapprobation, attempt to discourage him, before I had examined and reflected upon it. By the examination of many speculative subjects, abounding with falsehood, we are frequently enabled to treasure up some truths. Some of the first and most important discoveries in chemistry, owe their origin to the midnight vigils of the alchymists, who vainly sought for the philosopher's stone: and many valuable combinations in the science of mechanics have been discovered by those who wasted years in as vain a pursuit, after a perpetual motion.

I believe there are but few theories, which do not contain much that is profitable. The man who has the ingenuity to advance new ones, will be likely, in the course of reasoning necessary to support them, to say something that is useful to be known. In his very reveries and wanderings, he will often point out land-marks, which may be useful to the future *traveller*. Whether then is it better to crouch under the tyranny of prejudice, or employ our thoughts and reasoning powers in the search of truth, though at the risk of deceiving ourselves, as our predecessors have done? Had it not been for a prudent boldness in advancing and defending new doctrines, the human mind must have remained to this day, the sport of all the chimeras of the ancients.

The exact shape and formation of the earth are admitted not to be well understood. The laws of gravity, and the admeasurements which have been made in different places on the same meridian, have demonstrated to us, that the greatest mathematicians have mistaken its real figure. Various theories have at different times been published and refuted, and others substituted in their stead. Yet still a shade of darkness and mystery appears to hang over the subject; for many principles, attractions, and apparent variations from the established laws believed to exist in the economy of nature, have been discovered, particularly in the polar regions, which remain unexplained and unaccounted for. Let us, therefore, examine and investigate any theory which proposes to explain them. Let us not be so tenacious of our own opinions, and hereditary prejudices, as to stop at the very point where every thing invites us to proceed. Let us rather push our researches after knowledge to the utmost, and exercise our reason, and every means in our power that may tend to the advancement of science and knowledge. In the pursuit, let us not be retarded by the cry of prejudice, or the sarcastic whispers of the narrow minded, and selfish.

Let us, therefore give Captain Symmes a "gentle meeting," and a candid hearing, in the following short chapters; ascertain what his theory is, and on what principles he supports it; and then adopt or reject it, as our reason may dictate.

# CHAPTER II.

*Symmes's Theory; comprehending his description of the form of the earth, and of the other orbs in the Universe; his principles of gravity, and the points wherein he differs from the old or generally received theories.*

According to Symmes's Theory, the earth, as well as all the celestial orbicular bodies existing in the universe, visible and invisible, which partake in any degree of a planetary nature, from the greatest to the smallest, from the sun, down to the most minute blazing meteor or falling star, are all constituted in a greater or less degree, of a collection of spheres, more or less solid, concentric with each other, and more or less open at their poles; each sphere being separated from its adjoining compeers by space replete with aerial fluids; that every portion of infinite space, except what is occupied by spheres, is filled with an aerial elastic fluid, more subtile than common atmospheric air; and constituted of innumerable small concentric spheres, too minute to be visible to the organ of sight assisted by the most perfect microscope, and so elastic that they continually press on each other, and change their relative situations as often as the position of any piece of matter in space may change its position: thus causing a universal pressure, which is weakened by the intervention of other bodies in proportion to the subtended angle of distance and dimension; necessarily causing the body to move towards the points of decreased pressure.

It is a sound principle of philosophy, that the particles of the common air of our atmosphere are of a repellant quality, and mutually repulse each other. The whole system of pneumatics goes to prove that air presses equally in all directions. Not a single experiment in this branch of natural science can be performed that does not depend on such a property. This being the case, if the boundless extent of the universe, beyond the limits of our atmosphere, be an entire vacuum, why should the atmosphere be retained in its present circumscribed form, and not expand, by virtue of its repellant quality, far beyond its known height? To prevent this, Symmes believes universal space to be filled with an elastic fluid, inconceivably rare, and uniformly distributed throughout; differing from common air, and from the elastic fluids (which also are known to be repellant) existing in our atmosphere. This tendency is what Symmes believes should be understood by the term gravity; the laws of action governing which he holds to be true, as defined by Newton: and he moreover holds that the application of the laws of gravity, as laid down by Newton, leads a reasoning mind to the belief of concentric spheres, with open poles, as all planetary bodies are in his opinion formed.

In regard to the *effects* of gravity, he pretends not to differ from the generally received opinion of the age; but the *application* of them, as to the inner parts of insulated bodies, has enabled him to *improve* in a knowledge of the formation of planets; and finally led him to form a correct idea of what *constitutes* gravity.

The author of the new theory entertains a belief that the principles of planetary orbicular forms, developed by him, extend as well to the molecules of the most subtile fluids, as to the innumerable stars or suns of the universe, and all their planetary trains: he contends that though he may not have discovered any new principles in physics, yet that he has made interesting advances in a knowledge of the application of what was heretofore known.

According to him, the planet which has been designated the Earth, is composed of at least five hollow concentric spheres, with spaces between each, an atmosphere surrounding each; and habitable as well upon the concave as the convex surface. Each of these spheres are widely open at their poles. The north polar opening of the sphere we inhabit, is believed to be about four thousand miles in diameter, and the southern above six thousand.[1] The planes of these polar openings are inclined to the plane of the ecliptic at an angle of about twenty degrees; so that the real axis of the earth, being perpendicular to the plane of the equator, will form an angle of twelve degrees with a line passing through the sphere at right angles with the plane of the polar openings; consequently the verge of the polar openings must approach several degrees nearer to the equator on one side than on the other. The highest north point, or where the distance is greatest from the equator to the verge of the opening in the northern hemisphere, will be found either in the northern sea, near the coast of Lapland, on a meridian passing through Spitsbergen, in about latitude sixty-eight degrees, or somewhat more eastwardly in Lapland; and the verge would become *apparent*, to the navigator proceeding north, in about latitude ninety degrees.

The lowermost point, or the place where the distance is least from the equator to the verge of the northern polar opening, will be found in the Pacific ocean, about latitude fifty degrees, near the north-west coast of America, on or near a meridian running through the mouth of Cook's river, being in about one hundred and sixty degrees west longitude, the real verge being in about latitude fifty degrees and becoming apparent to a person travelling northward at right angles with the magnetic equator, at the distance of about twelve hundred miles further. The verge varies progressively from the lowest to the highest point, crossing the north-west coast of America between latitude fifty-two and fifty-four, thence across the continent of North America, passing through Hudson's Bay and Greenland, near cape Farewell; thence by mount Hecla to the highest point; thence tending gradually more to the south, across the northern parts of Asia, at or near the volcanoes of Kamtschatka, and along the extinguished volcanoes of the Fox Islands, to the lowermost point again, near the northwest coast.

In the southern hemisphere, the highest point, or place where the distance is greatest from the equator to the verge of the polar opening, will be found in the southern Pacific ocean, in about latitude forty-six degrees south, and perhaps about longitude one hundred and thirty degrees west; and the lowermost point, or place where the distance is least from the equator to the verge of the opening, will be found on a meridian south or south-east of the island of Madagascar, in about latitude thirty-four degrees south, and longitude about fifty degrees east; thence passing near the cape of Good Hope, across the Atlantic ocean, and southern part of the continent of America, through a chain of active volcanoes, to the highest point; thence bearing regularly toward the lowest point, passing between the two islands of New-Zealand, or across the most southerly one, and the northernmost part of Van Dieman's land, to the lowest point, which is south or south-east of Madagascar; the apparent verge being several hundred miles beyond the real verge.[2] Consequently, according to this formation of the sphere, the degrees of latitude, on different meridians, will vary according to their distance from the polar openings; and the magnetical equator, which encircles the sphere, parallel to the plane of the polar openings, would cut the real equator at an angle of twelve degrees. A person standing on the highest part of the apparent verge would appear to be under the polar star, or nearly so, and at the ninetieth degree of latitude. The meridians all converge to the highest point of the verge, or the ninetieth degree; consequently, in tracing a meridian of longitude, you would pursue a direction at right angles to the equator, until you arrived in the neighbourhood of the real verge of the polar opening, when the meridians would change their direction and turn along between the real and apparent verges towards the highest point, until they all terminated at the ninetieth degree of latitude; this being the direction a person would travel in order to have his back to the sun always at 12 o'clock, the time of his greatest altitude. Although the particular location of the places where the verges of the polar openings are believed to exist, may not have been ascertained with absolute certainty, yet they are believed to be nearly correct; their localities having been ascertained from appearances that exist in those regions; such as a belt or zone surrounding the globe where trees and other vegetation (except moss) do not grow; the tides of the ocean flowing in different directions, and appearing to meet; the existence of volcanoes; the "*ground swells*" in the sea being more frequent; the Aurora Borealis appearing to the southward; and various other phenomena existing in and about the same regions, mark the relative position of the real verges.

The heat and cold of the different climates are governed by their distance from the verge of the polar opening, and do not depend on their nearness to or remoteness from the equator. The natural climates are parallel to the planes of the polar openings, and cut the parallels of latitude at an angle of twelve degrees. When the sun is on the tropic of Capricorn, the circle of greatest cold would be about twenty-three and a half degrees south of the apparent verge, and when the sun is on the tropic of Cancer this circle would probably be just under the umbrage of the real verge: hence it follows, if this doctrine be correct, that the climate of forty degrees north latitude on the plains of Missouri, in the western part of the continent of America, will be as cold in winter, as the latitude of fifty or fifty-two degrees in Europe; and observation has fully confirmed such to be the fact.

The magnetic principle which gives polarity to the needle, is believed to be regulated by the polar openings, and that the needle always points directly to the opening, and of course parallel to a line drawn perpendicular to the plane of the opening. And when the apparent verge shall be passed, the needle will seem to turn nearly round, so as to point in an opposite direction; having the contrary end north on the interior of the sphere, that was north on the exterior, the same end being north on the interior which was south on the exterior. Hence, when navigators arrive in the neighbourhood of the apparent verge, the variation of the needle becomes extreme; and when the verge is passed, the variation is more or less reversed. The meridians run from the highest northern to the highest southern point on the verges; hence, in tracing a meridian, or sailing due north, we would pursue that line which would conduct us directly from the sun at his greatest altitude; and when we come to the verge, the meridian would vary, and wind along the vicinity of the edge of the real verge, until it brought us to the highest point of the apparent verge. The magnetic needle, on arriving at the verge would appear to cease to pursue the same direction, but would in reality continue to maintain it, and lead directly into the polar opening.

According to this formation of the sphere, a traveller or navigator might proceed true north any where west of the highest point of the verge, say on the continent of America, until he come to the verge. The meridian on which he was travelling would then wind along the verge to the right, until he arrived at the ninetieth degree; and by proceeding south, in the same direction, he would arrive at the coast of Siberia, without going far into the concavity of the sphere, and without knowing that he had been within the verge. Should such a journey be effected, it would appear to confirm the old theory of the form of the earth, and put the subject at rest; although pursuing the needle might have directed the traveller into the interior, and enabled him to discover those fine countries which Captain Symmes alleges to exist there.

Each of the spheres composing the earth, as well as those constituting the other planets throughout the universe, is believed to be habitable both on the inner and outer surface; and lighted and warmed according to those general laws which communicate light and heat to every part of the universe. The light may not, indeed, be so bright, nor the heat so intense, as is indicated in high northern latitudes (about where the verge is supposed to commence) by the paleness of the sun, and darkness of the sky; facts, which various navigators who have visited those regions confirm; yet they are no doubt sufficiently lighted and warmed to promote the propagation and support of animal and vegetable life.

The different spheres constituting our planet, and the other orbs in creation, most probably do not revolve on axes, parallel to each other, nor perform their revolutions in the same periods of time; as is indicated by the spots on the belts of Jupiter, which move faster on one belt than another.

The atmosphere surrounding the sphere is probably more dense on the interior than the exterior surface, the increased pressure of which must increase the force of gravity; as the power of gravity must increase in proportion as we approach nearer the poles.—Clouds formed in the atmosphere of the convexity of the sphere, probably float in through the polar openings, and visit the interior, in the form of rain and snow. And the long continuation of winds, or regular monsoons, which occur in some parts of the earth, may be supplied by winds sucked into one polar opening and discharged through the other, thus performing the circuit of the sphere; without which supposition, it would be difficult to account for the long continued winds which, at certain seasons, are known to blow constantly for several months, more or less obliquely to and from the poles.

The disciples of Symmes believe that each sphere has a cavity, or *mid-plane space* near the centre of the matter composing it, filled with a very light, subtile, elastic substance, partaking somewhat, perhaps, of the nature of hydrogen gas; which aerial fluid is composed of *molecules* greatly rarified in comparison with the gravity of the extended or exposed surfaces of the sphere. This *mid-plane space* tends to give the sphere a degree of lightness and buoyancy. Besides this large *mid-plane space*, perhaps numerous other interstices exist in the sphere nearer the surface, and of more limited extent. The gas escaping from these spaces is, no doubt, the cause of earthquakes;

and supply the numerous volcanoes. This gas becoming rarified and escaping, must occasion most of those great revolutions and phenomena in nature, which we know to have occurred in the geology of the earth. This aerial fluid with which the *mid-plane spaces* are filled, may possibly be adapted to the support of animal life; and the interior surfaces of the spheres formed by them, may abound with animals, with organs only adapted to the medium which they are destined to inhabit.

In many parts of the unfathomable ocean there may be communications or passages from the surface of the sphere on the outer side to the surface of the inner, at least all except the great *mid-plane space*, through which liquid apertures, light and heat may be communicated, perhaps, to the interior surface of the sphere.

# CHAPTER III.

*Symmes's Theory supported by arguments drawn from the principles inherent in matter, and the consequences resulting from motion; tending to show that, from necessity, matter must form itself into concentric circles or spheres, such as Symmes describes the earth to be composed of.*

It is a principle laid down by Sir Isaac Newton, the correctness of which is generally admitted, that "matter attracts matter in proportion to its quantity and the squares of its distances inversely." Captain Symmes contends that gravity consists in a certain expansive quality in the molecules which constitute the aerial fluid called æther, which fills universal space, and creates a pushing, instead of a pulling power. However, let either be correct, I conceive it cannot materially affect the principles necessary to constitute concentric spheres: either principle, I apprehend, would lead us nearly to the same results. When matter was in chaos, or in a form not solid, promiscuously disseminated through universal space, suppose it then should at once receive the impression of those universal laws by which it is governed, and see what would be the consequence.

According to Sir Isaac Newton's principles of gravity, the particle of matter that happened to be the largest would attract the smaller in its neighbourhood, which would increase the power of attraction in proportion to the increase of matter, until all in the universe would be collected into one vast body in the centre of space, and there remain motionless and at rest forever. This, however, we find not to be the case; for innumerable bodies of matter, differing in magnitude, are known to exist throughout the universe, arranged at suitable distances from each other, and performing certain revolutions in obedience to certain fixed laws impressed on them.

Now suppose all the matter in our globe to be an extended liquid mass, the particles so disengaged from each other, as to take their positions according to the *established* laws of matter, and then see what would be the consequences resulting from motion and gravity. Taking the laws of Newton for our guide, the particles of matter in the centre would be operated on by the power of gravity equally on all sides and consequently be stationary. Suppose then a line struck through this globe of matter, so as to make a globe of half the diameter of the whole in the centre, it is plain that the inner globe would not contain more than one eighth part as much matter as the surrounding one; hence it would be attracted towards the surface more than to the centre, were it not for the attraction of the matter on the opposite side exerting an influence upon it—but this being removed to so much greater distance, would not be more than an equipoise to the other.

The diameter of our globe, according to the best observation, is believed to be about 7970 English miles, and its circumference 25,038: consequently, if it were solid, it would contain 265,078,559,622 cubic miles of matter; while a globe of only half the diameter, would contain only 33,134,819,952.[3]

Suppose our globe divided into parts of one square mile on the surface, bounded by straight lines converging to a point at the centre, as the subjoined figure represents:

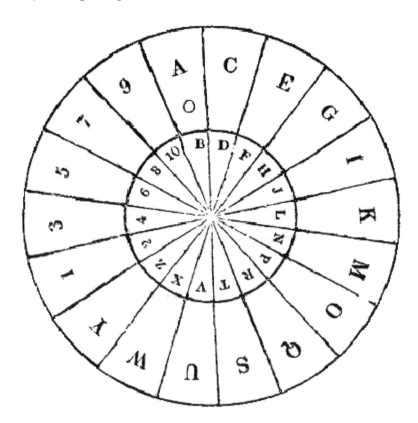

and then suppose there were no other particles of matter in the universe but A and B, A containing 1,328 cubic miles of matter, and B only 166, A would attract B so as to make their centre of attraction at O, which point would become at once the common centre: but admitting the whole matter of the globe to exist, A would still exert its influence on B, but both would be operated upon by T and S and the surrounding matter, all perhaps, tending to one common centre. However, I imagine that the tending to the centre would not be so great as is contended for by the generally received theory, which alleges that matter at the centre of the earth is four times as hard as hammered iron. The Newtonian philosophy appears to contemplate a globe at rest, and not in such rapid motion as we know the earth and other planetary bodies to be in, communicating to them a centrifugal force, which tends to throw matter from the centre. The rotary motion of each planet is no doubt regulated by the quantity of matter it contains: so that at its surface centrifugal and centripetal forces are equally balanced—the rotary motion being adequate to communicate a force to counter-balance the force of gravity.

Newton ascertained by his investigations of the properties and principles of matter, the earth to be a globe flattened at the poles: and the French philosophers afterwards confirmed this fact by measuring a degree in different latitudes. This difference between the equatorial and polar diameters of the earth, and of the other planets which are also known to be of that shape, is ascribed by those philosophers who attempt to account for such a formation, to the projectile force of the globe at the equator occasioned by its rotary motion. This is admitting that the matter of our globe was once in so soft a state as to take its form from motion; for were the earth a compact solid body, and four times as hard as hammered iron at the centre, (as the Newtonian system alleges) this rotary motion round an imaginary axis could never give to the globe the form of an oblate spheroid, as is ascertained to be the fact; because a hard solid body moving in empty space, could not be supposed to yield into that shape by any law of action as yet unfolded by science.

But were the matter of this globe thrown into a confused, disorganized state, and then put into a quick rotary motion, such as it is known to have, it would throw off from the centre towards the surface, first the heaviest, and next the lighter substances, which is the very order in which they are found to be arranged, in the composition of the earth.

This principle, for it is simply the principle of projectile force, will account for mountains, hills, vallies, plains; and for nearly all the inequalities on the face of the earth. These circumstances depend on the density of substances composing the earth. Substances of the greatest specific gravity are susceptible of the greatest projectile force; and hence we find that mountains are composed of heavy masses of rock, mineral substances, and heavy earths; hills, or the next highest eminences, of earth of the next specific gravity; and plains, or level lands, of lighter substances. Had the earth originally been composed of one uniform substance, sand, for example, of equal fineness and weight, the whole surface of the globe would have presented one uniform level or unbroken plain. But, presuming that it was originally composed of, at least, earths of different densities, the heaviest masses would be first thrown out and raise their heads above the surface of the ocean: thus islands would be formed; and clusters of islands would form continents, rearing their lofty heads into the air; and, if the substances of which they were originally composed, were not as hard as the rocks which we now find on them, the sun and changing temperature of the climates, might convert certain kinds of earth into masses of stone, increasing in specific gravity by petrifaction, and other causes, until the towering peaks of the Alps and Andes assumed their present solid form. One continent having thus emerged, another would naturally be produced simultaneously on the opposite side of the sphere, as an equipoise to the first, to keep equal the earth's motion; until all the heavy substances should be thrown out and united in a compact sphere.

To an observer of the earth the crust every where appears to indicate the emergence of land from water: almost the whole surface of the solid crust is alluvial, and by reasoning and reflecting, we are led to the conclusion, that the solid parts of our globe are nothing more than a crust, and formed into concentric spheres, in accordance with the principles of projectile force. I would ask, what proofs have we, that the sphere we inhabit is solid beyond the degree of thickness necessary to preserve it from injury by its rapid motion round the sun, by its diurnal motion round its own axis, and by its motion round its common centre of gravity with the moon? It has been ascertained with mathematical certainty, that the large planet Jupiter, is more than 1300 times the bulk of the earth, and Saturn independent of his double ring, is about 1000 times the size. If we apply to those prodigious bodies, the reasoning of Newton relative to plastic forms moving variously, there is no just grounds for concluding that they are solid substances to their centres. If they were, their vast weight and remote position would require much more attraction than probably even the sun could furnish, to keep them within their orbits.

The acknowledged and received laws of gravity, together with the measurements made on the same meridian, in different latitudes, have demonstrated to us that the greatest mathematicians have been mistaken as to the real figure of the earth. It is for schoolmen to make exact calculations, respecting the force of gravity, and centrifugal and centripetal forces; it is for them to determine with mathematical certainty where matter, left to its own laws, would settle; for such undertakings, I acknowledge my incompetency. But I have long had strong doubts, whether the laws of gravity are well understood; or whether the rules on which these calculations could be accurately made, are exactly known.

However, I take the broad principles of nature, as presented to my view, for my guide; and draw my conclusions from what I have seen or what is well known to exist.

Observe the boy hurling a stone from a sling; he whirls it round his head for a minute to acquire a certain degree of centrifugal force, and although it is not whirled with half the velocity the earth revolves on its axis, yet as soon as it is released from confinement, notwithstanding the whole power of the earth is operating on it with all the force of gravity, the centrifugal force which the stone acquired by the whirling is sufficient to carry it off, at a tangent to the circle described by the sling, for a very considerable distance, before the gravity of the earth and atmospheric obstruction can force it to the ground.

If you will take the trouble to examine a mechanic grinding cutlery on a large stone that is smooth on the sides and has a quick motion, you may observe that if a certain portion of water be poured on the perpendicular side whilst the stone is turning, it does not settle or form itself into a body round the crank or axis; nor does the gravity of the earth draw it from the surface, but forms itself on the side of the stone into something resembling concentric circles, one within another. The surface of the earth, I apprehend, revolves with much greater velocity than any grindstone; and the substances composing the spheres are much firmer than water.

Most of us, I presume, have seen persons for amusement, in displaying feats of dexterity, place a full glass of wine or water on a hoop, and whirl it round their heads without spilling one drop. The centrifugal force it acquires by the revolutions overcomes the power of gravity, although nothing appears to support it but the common atmosphere.

Another experiment, producing a similar effect, might be made with a cup filled with fine sand. On the surface of the sand, describe a circle nearly in the centre; it will then be apparent, on observing the cup, that the sand within the circle, provided the particles attract one another as the planets do, is as much attracted towards one verge of the cup as the other; owing to its being equally surrounded by matter or sand, and therefore it can be but very little, if any, gravitated centrewise. Hence, being in a degree suspended, only a small horizontal rotary motion is required to whirl it towards the rim or sides of the cup into a circular form; and hence it follows, that those particles of sand lying equidistant from the inner side of the circle of sand thus formed, and the outer side would be in like manner balanced, or supported, by being equally gravitated in both directions. A disposition would thus be produced to form into concentric circles, and it would therefore follow, that successive similar dispositions to subdivision should occur, gradually lessening in force and quantity. This principle applied to the earth or other planets, would cause them to be formed into concentric spheres; and would throw the matter from the axis, as well at the poles, as at the centre, and thereby constitute open poles.

Another simple experiment might also be made, to illustrate that a disposition to concentric spheres does exist in nature. On a piece of paper sift a small quantity of very fine magnetic particles, such as steel or iron filings, under which hold a loadstone; and you will observe that the attractive power of the magnet will cause the filings on the paper to arrange themselves into various concentric circles, nearly regular and equidistant from each other. From what cause should this take place, rather than that the filings should be accumulated into one mass?

Various have been the conjectures relative to the cause and origin of the meteoric stones, or fire balls, which have been known to fall to the earth, in all ages, and in various parts of the world. Some have imagined them to be precipitated from a comet or some of the planets; others that they come from the moon; and Captain Symmes's opinion, I believe, is that they are formed isolated in space by spontaneous accumulations, as by attracting molecules of matter at first in a fluid state, which afterwards solidifies by heat or motion. But come from whence they may, they are said to be constituted of a substance unknown to our geologists; and in several instances the fragments have been ascertained to consist of pieces, some of which have concave and some convex surfaces, affording a certain proof that previous to their descent, they had been constituted of hollow spheres.

Professor Silliman, of Yale college, has preserved some of the fragments of one of these fire balls; and in his valuable journal, has given the public an able description of the facts which occurred, when they fell. This fire ball fell in the state of Connecticut, in the year 1807, producing three distinct reports, like a cannon, making three convulsive leaps or throes in its course, which were simultaneous no doubt with the explosions, becoming less luminous after each, and being quite extinguished at the third. Three showers of stones fell to the earth in a line with its course; the second shower fell five miles distant from the first, and the last three or four miles from the second. Some of the fragments were found to be concave, others convex, and especially on those sides of the fragments which were glazed with sooty crusted surface, as if vitrified.

These phenomena are precisely such as would occur, supposing the fire ball to have been a small satellite, or erratic planet, at first fluid, which had become so condensed by the increased action of terrestrial gravity, occasioned by its sudden approach, as to cause its fluid parts to chrystalize and form into, at least, three concentric spheres; and the latent heat and light set free by such rapid condensation as to produce the meteoric flame; which in this case was almost equal in light to that of the sun at mid-day. As soon as the spheres became sufficiently solidified to prevent the heated aerial fluid, contained in the mid-plane cavities of the spheres, from passing out with freedom, when expanded by the heat; or let the atmospheric air pass in, in case a condensation within afforded a vacuum; the solid crusts of the spheres would be disruptured successively one after the other; lose their regular rotation, and fall in fragments to the earth. The fall of this body is not a solitary instance of the kind: others have fallen in many parts of the earth, attended with phenomena more or less the same.

On the 16th of January, 1818, in Florida, near Mobile bay, a fire ball bursted with a considerable report. Immediately before the explosion, it was observed to project a cone of fire from each pole horizontally and at right angles with its course. Its bursting like a bomb-shell, indicated that it must have been hollow; and the two cones of light which appeared, beside its train, showed that it was open at the poles.

Turn your attention to the general economy of nature throughout her works, and you will perceive in various and almost innumerable substances that she forms hollow cylinders or spheres in the room of solid ones. Enquire of the botanist, and he will tell you that the plants which spring up spontaneously, agreeable to the established laws of nature, are hollow cylinders. If a hollow globe would answer the ends of supporting organic life as well as a solid one—why not be hollow, as well as a stalk of wheat? or by what laws is the stalk of wheat governed, that it should *always* grow hollow? What law in nature causes the quills and feathers of a bird to be hollow cylinders? Why are they not solid? I presume it is for this plain reason, that nature, throughout all her works, has wisely assigned to every thing just matter enough for strength and usefulness; and has in no case overburthened it with unnecessary and cumbrous weight.

Enquire of the anatomist, and he will tell you that the large bones of all animals are hollow, and particularly that the bones of birds are more than ordinarily so: even the minutest hairs of our heads are hollow.

Go to the mineralist, and he will inform you that the stone called Ærolites, and many other mineral bodies, are composed of hollow concentric circles; and, that strata of different kinds abound in various mineral substances. Even the earth itself is composed, as geologists tell us, of various strata, composed of different substances, and varying from one degree of density to another. If every part of our globe be regulated according to the received laws of gravity, and the relative density of matter, why do we find almost all over the world, light alluvial soil in the vallies and plains; and on the tops of the highest mountains, the more heavy granite, and some of the heaviest substances that nature knows? We can hardly indulge the thought that all this is the work of volcanic eruptions or some dread throe of nature.

However, if we direct our attention alone to those general laws which are known, and which are believed to govern matter, I apprehend it would be very difficult to account for the creation of worlds, and the admirable arrangement which subsists throughout the universe. To account for every thing, either according to the old or new theory, would be attempting too much. It would be placing the Deity in some corner of the universe an idle spectator, whilst matter governed by its own laws, was forming itself into worlds and systems; the bare thought of which is irreverent. Is the existence of matter owing to some other first cause, or did matter create itself, and impress upon itself the laws which govern it? Such an idea is absurd. We might as well imagine that matter created God, as itself. By attempting to trace every effect to some natural cause, is attempting to do more than we shall ever be able to accomplish. Such a course of reasoning must lead us to the conclusion that there is no God, or first cause; or, at least, to what would be nearly the same thing, that there is no need of one.

But in reasoning upon this subject, I take it for granted, that there is a God, and that he is the first cause of all things, the creator of all the orbs in the universe, be they either solid globes or concentric spheres; and I hope such is the reader's belief. And I cannot discover in this any thing derogatory from His infinite power, wisdom, or divine economy, in the formation of a hollow world and concentric spheres, any more than in that of solid ones. I should rather be of opinion, that a construction of all the orbs in creation, on a plan corresponding with Symmes's theory, would display the highest possible degree of perfection, wisdom, and goodness—the most perfect system of creative economy—and, (as Dr. Mitchill expresses it) *a great saving of stuff.*

# CHAPTER IV.

*Arguments in support of Symmes's Theory, drawn from celestial appearances.*

That a disposition to hollow cylinders does exist in nature, I think, must be admitted; and that a similar principle exists in the planetary system, at least in some degree, appears to me as certain. Every person has seen or heard of Saturn and his rings. At certain periods of time the appearance of this planet, viewed through a good telescope, represents him to be surrounded with two luminous rings or bodies of matter, concentric with each other, and with the body of the planet. These rings no where adhere to the body of the planet, but are distinct and separate, some considerable distance from him, and from each other, leaving a portion of vacant space between the planet and the rings, through which we see the fixed stars beyond.[4] It is a fact, I believe, admitted by all, and of which we have positive ocular demonstration, that these rings are constituted of some kind of matter, if not solid, at least to all appearance as much so as the body of the planet. Their thickness must be very inconsiderable, for when the edge is turned to the eye it is no longer visible, except to the powerful reflecting telescope of Dr. Herschel.—Thus the rings undergo phases according to the position of the planet in his orbit, which prove them to be opaque, like other bodies in the planetary system, and like them shining by reflection. I am not informed what is the precise velocity of the rotary motion of the rings; probably their varying aspect, or some other cause has prevented a correct observation from being made. However, the planet itself revolves on its axis, with an astonishing velocity; and no doubt the rings also, though perhaps with different degrees of velocity.

The appearance of Saturn, I conceive, establishes the fact, that the principle of concentric spheres, or hollow planets, does exist, at least in one instance, in the solar system. And if the fact be established that it exists in one case, is it not fair, nay, is it not almost a certain and necessary consequence, that the same laws of matter which formed one planet into concentric spheres, must form all the others on a plan more or less the same? If we draw any conclusion, or form any opinion at all, respecting the formation of the planets, whose inner parts we cannot see; or if we form any opinion in relation to our own planet in particular, whose poles have never been explored, would not reasoning from analogy bring us to the conclusion, that all bodies of matter are formed similar to that of Saturn, unless we have positive proof to the contrary? But it is not in Saturn alone that we find proof of the principles contended for by Captain Symmes. Most, if not all of the other planets, belonging to our system, whose relative situation afford us an opportunity of observation, appear to exhibit strong proofs that the same principles prevail throughout.

The planet Mars, exhibits concentric circles round one or the other of his poles, according as either is more or less in opposition to us. These circles appear alternately light and dark, exactly as they should, supposing the planet to be constituted of concentric spheres, (such as Symmes believes of the earth) the light being reflected from their verges on which it falls; and in which case the vacant space between the spheres would necessarily appear dark.

Sometimes he appears to us with a single ring at each pole. At such times his axis is at right angles, or nearly so, with a line drawn from the earth to his centre. This, I conceive, can be accounted for by the great refraction, occasioned by the increased density of his atmosphere around the poles, which appears to throw out the further sides of the verges so as to make them appear like rings, in the form they present themselves to our view. That such is the natural appearance may be evidenced by taking a small wooden sphere with open poles, and immerse it in a circular glass vessel filled with water; when viewed horizontally through the side of the glass, with the plane of the openings at a right angle with the visual ray, the refraction occasioned by the water, answering to the dense atmosphere of Mars, will apparently throw out the polar openings, and present you with a view, similar to the appearance of Mars, when his axis is at right angles to us.

Our next neighbour, Venus, between us and the sun, (though her being between us and the sun prevents us from having so favourable an opportunity of examining her poles, as those of Mars, who is our next neighbour on the side opposite the sun) presents appearances at certain times, which seem to lead to the conclusion, that she also is constituted of concentric spheres. At times, when this planet is nearly a crescent, we are able to discover a deficient space near the tip of one of her horns. Admitting Venus to be constituted of concentric spheres with open poles; and supposing one of the vacant spaces, between two of her spheres about the polar openings, to traverse her horn or cusp, at the place where the dark space occurs,—it would present to us exactly such an appearance as does actually occur.

At other times, one of the horns or cusps of Venus is seen to wind inward as it were into the body of the planets, extending about fifteen degrees further than the other horn. This is an appearance which would also be presented, if Venus is formed according to Symmes's theory. And again, supposing one of her horns to terminate around the verge of a polar opening, in such way as to follow the curve of the verge for some distance, (which is of course more curved than the periphery of the planet) and the same appearances, I think, would occur. The axis of the planet not being at right angles with the polar openings, in its revolutions one side of the verge would be thrown much nearer to us than the other; and the different spheres revolving on their axes with different velocities would at different times exhibit to our view the verge of a different sphere.[5]

The axis of the planet Jupiter is always at right angles with a line drawn to the earth, consequently his poles are never presented to us; but his belts, which we can and do see, seem to speak loudly in favour of a plurality of spheres. The most common appearance of Jupiter is, that he is surrounded by four belts; two bright and two dark, alternate to each other. But they are variable, presenting different appearances; at some times seven or eight belts are discoverable, at other times they appear interrupted in their length, and to increase and diminish alternately, running into each other, and again to separate into a number of belts of a smaller size. If Jupiter be a solid globe, I would enquire, how is it possible to account for those various changes in his belts, or even for their existence at all? Astronomers, I understand, have heretofore considered the phenomena of Jupiter's belts as altogether unaccountable. If he be a simple plain globe, those belts could not exist; or if they did, they must forever remain uniform, and not change their size and shape, or relative positions in respect to each other; neither could the spots on one belt rotate faster than those on another. But if we adopt the doctrine of concentric spheres, and that this planet is composed of a number of them, we can account at once for all the various appearances in a rational manner. The belts would be produced by the shadow cast on the space between the polar opening of one sphere and the adjoining one; that is, a portion of the sunshine, would be reflected from the verges of the spheres on which it fell; and another portion would appear to be swallowed by the intervening space. And if refraction bends the rays of vision between and under his spheres, as it bends a portion of the rays of the sun, so as to produce the apparent belts of comparative shade, then a very complete solution of those appearances, heretofore considered wonderful, would be afforded. The variation which has been observed in their number, shape, and dimensions, can in no way be better accounted for, than by concluding the planet to be constituted of a number of concentric spheres, of different breadths, revolving on different axes, and with different velocities, so as sometimes to present to our view the verge of one sphere, and sometimes that of another: and the rays of the sun falling on the parts of the verges presented to us, would occasion the diversified appearances which we discover. If some sections of both crusts of the spheres be formed of water alone, and become occasionally transparent, it will afford an additional reason for the varying phenomena attendant on these appearances, which may also be increased by alternate regions of water, ice, dry land, and snow.

Modern astronomers have long noticed the spots frequently visible on the sun. They are described as having the appearance of vast holes, or fractures, in his outer surface or crust, through which an inner appears to be seen. This, also, seems to favour the doctrine of different spheres. Notwithstanding the sun revolves very slowly on his axis, it is probable that his poles are open to a greater or less extent; but we can never see into them, owing perhaps to the earth, never being very far from the plane of the sun's equator, his being such a vast deal larger than the earth, and the atmosphere surrounding him so extremely luminous.

Very little doubt exists in my mind, that the poles of the sun and of Jupiter would appear somewhat like those of Mars or the rings of Saturn, were it not that the two former never present their axes, in any perceptible degree, towards us; neither does our satellite, the moon, ever present either of her poles to us: hence, though this may be in some degree open, (notwithstanding her slow rotation) owing to her axis always being nearly at right angles with a line drawn to the earth, we are not able to see whether they are open or not,—more especially as her atmosphere is so light and rare as not to produce much refraction. The vast round deep caverns observable on the surface of the moon, appear as if they might once have been polar openings; if so, she must frequently have changed her axis.

The spots of light which have at different periods been discovered by astronomers, on the surface of the moon, near her poles, when she was on the face of the sun, in an eclipse of that luminary, are perhaps best accounted for by supposing the sun to shine in, either at one of her polar openings or through a cavity on her further side, and appearing to us through one of her annular cavities, on this side, and near her poles: Or the sun being much larger than the moon, and the axis of the moon a little varied from right angles with the earth, (or perhaps the low side of the sphere being next to the earth,) the sun would shine through an annular cavity or open pole, so as to appear to us as a spot of light on the moon's disk.

The foregoing enumerated astronomical phenomena are some of the facts tending to confirm and elucidate Symmes's theory. They all have been long known to exist; yet I have never heard them accounted for to the satisfaction of my mind. Indeed, I believe some of them never was attempted to be accounted for in any manner whatever. I would, therefore, request the reader, who may deign to give the subject a serious thought, to reflect, that if all the celestial orbs are entire round globes, as the old theory considers them to be, on what principles, or in what manner, could they present the various appearances which I have enumerated? Why should the horns of Venus assume different shapes? What would make the appearance of belts on Jupiter? Or rings and concentric circles at the poles of Mars? And, finally, in what position could a round solid globe be placed, to exhibit the rings of Saturn, revolving with different velocities, as it respects each other, and spaces appearing between them and the body of the planet, through which stars, millions of miles beyond, can be distinctly seen? These are phenomena I should like to hear explained. On the principle of concentric spheres, they can all be accounted for in a most satisfactory manner. They appear perfectly plain and intelligible. What was thought to be involved in inexplicable mystery, and mid-night darkness now perfectly accords with the established laws of nature, and can be understood by the most ordinary capacity.

# CHAPTER V.

*The Theory of Concentric Spheres, supported by arguments drawn from Terrestrial facts; such as the migration of animals to and from the arctic regions, and from refraction, and the variation of the compass, observed in high northern latitudes.*

I would now advert to a few of the known terrestrial facts, which have a tendency to support the theory advanced by Captain Symmes; such as the migration of animals, including beasts, birds, and fishes, in the arctic regions; and from refraction, and the variation of the compass observed in high northern latitudes.

It is a fact well attested by whalers and fishers in the northern seas; and one that almost every author who adverts to the northern fisheries confirms, that innumerable and almost incredible numbers of whales, mackerel, herring, and other migratory fish, annually come down in the spring season of the year, from the artic seas towards the equator. Some authors describe the shoals of herring alone, to be equal in surface to the island of Great Britain. Besides these, innumerable shoals of other fish also come down. These fish when they first come from the north in the spring, are in their best plight and fattest condition: but as the season advances, and they move on to the southward, they become poor; so much so, that by the time they get on the coast of France, or Spain, as fishermen say, they are scarce worth catching.

The history of the migratory fish affords strong grounds to conclude, that the shoals which come from the north, are like swarms of bees from the mother hive, never to return; particularly the herring and other small fish. They are not known to return in shoals: and it is doubted by some writers on the subject whether any of them ever return north again, or whether they are not entirely consumed by men, and by other fish.

Whalers and other fishermen who go to the north, generally prosecute their business in the seas between latitudes sixty and seventy degrees, where whales are most abundant. Pinkerton, in his voyages, states, that the Dutch, who at different periods got detained in the ice, and were compelled to winter in high northern latitudes, could find but few fish to subsist on during the winter: which proves that the migrating fish do not winter amongst, or on this side of the ice.—All these facts relative to fish, appear to be well authenticated. Now, were the earth a compact and solid spheroid, according to the old theory; and were the seas frozen nearly to the bottom at the poles, as we would be led to conclude, where could all those fish, that come down to us every spring, breed? or, if they even all returned in the autumn, and all the north were a sea that did not freeze even to the poles, it would require a great stretch of credulity to imagine where they could obtain food for the winter; or even if their source of food were inexhaustible, could the region of the pole afford space sufficient for their health, so as to migrate south in the spring? If the earth be not hollow, (or at least greatly concave about the poles) where could all those fish find room in winter? But on Symmes's plan, admitting the globe to be a hollow sphere, and the inner, or concave part, as habitable as without, (at least as habitable for fish) the whole matter is at once explained.

Whales, and various fish, delight in cold regions. According to Symmes's Theory, a zone at a short distance beyond the real verge of the sphere, (which constitutes the coldest part, or as he has thought proper to term it, "the icy circle,") commencing at the highest point, in about latitude sixty-eight degrees, in the northern sea, near Norway, thence gradually declining to about latitude fifty degrees in the Pacific ocean, which is the lowest point, and thence regularly round again to the highest point. A certain distance beyond this, and short of the apparent verge, this zone, or icy circle exists, which is believed to be the coldest region of the earth. After passing this, we would advance into the interior of the globe, and into a milder clime. In the interior region, it is contended, those immense shoals of fish are propagated and grow, which annually come out and afford us such an abundant supply: nor does it appear that the interior parts of the sphere are altogether forsaken by the fish in summer; for shoals of fat mackerel and herring come down from the north in autumn, as well as in the spring.

The seal, another animal found in cold regions, is also said to migrate north twice each year; going once beyond the icy circle to produce their young; and again to complete their growth, always returning remarkably fat—an evidence that they find something more than snow and ice to feed on in the country to which they migrate.

Numerous other facts of importance, relative to the migration of quadrupeds, are well authenticated by travellers and others: particularly that of the rein-deer. In Rees's Cyclopedia, under the head, "Hudson's Bay," it is stated, that the rein-deer are seen in the spring season of the year, about the month of March or April, coming down from the north, in droves of eight or ten thousand, and that they are known to return northward in the month of October, when the snow becomes deep. Hudson's Bay is situated between sixty and sixty-five degrees north latitude. We are informed by professor Adams of St. Petersburgh, that on the northern coast of Asia, every autumn the rein-deer start north-eastwardly from the river Lena, and return again in the spring, in good condition: the mouth of the river Lena is in about latitude seventy degrees north. This appears to me rather a mystery according to the old theory of the earth, for why should those deer when the cold commences, seek a colder climate, and a more sterile country? The inhospitable coast of Liberia and Hudson's Bay, in the gloom of a dark winter, I should suppose, would be cold enough, without their seeking to spend the winter among nothing but eternal mountains of ice at the pole; where nature must be robed in snows and crowned with storms.

Hearne, who travelled very high north and northwest on the continent of America, details various facts in his journal, which strongly corroborate Symmes's position. Some of the facts he attempts to explain agreeably to his own ideas, and others he considers inexplicable. Among a great collection of facts, he states, that large droves of *musk-oxen* abound within the arctic circle, few of which ever come so far south as the Hudsons-Bay factories. He mentions seeing in the course of one day, several herds of those animals, of seventy or eighty in a herd, in about latitude sixty-eight degrees. He states that the polar white bears are very rarely found by any of the Indians in winter; and that their winter retreats appear to be unknown;[6] that they are sometimes seen retiring towards the sea on the ice in autumn; and appear again in great numbers in the latter end of March, bringing their young with them.

Hearne also states, that the white or arctic foxes are, some years, remarkably plentiful; and always come from the north; that their numbers almost exceed credibility; that it is well known none of them ever migrate again to the northward; and that naturalists are at a loss to know where they originate.[7] He also mentions that all kinds of game, as well as fish, in those high latitudes, are at some seasons excessively plentiful, and at others extremely scarce.

These facts strongly corroborate the doctrine of a hollow sphere: otherwise, why should the rein-deer, and other animals, migrate north instead of south; as our Buffalo on the plains of Missouri do, when pressed with snow and cold weather? Instinct generally leads animals to fruitful and productive, rather than unproductive, regions; why then proceed north on the approach of winter, unless in expectation of finding a warmer climate, or, at least, a more mild and plentiful country, beyond the icy circle? Independent of the immense droves of rein-deer, great numbers of musk-oxen, white bears, and white foxes, spend their winters towards the north; which tends to establish the fact, that a considerable extent of land must exist in that quarter of the earth. This, however, would infringe on the space necessary to accommodate the vast quantities of fish which appear to be propagated in that region, if the old system were true.

If we were to judge of the internal surface of the sphere, by its animal productions,—admitting that those animals heretofore enumerated, are propagated there,—we should conclude that the internal region of the earth is as much more favourable to the support of animal life, as the rein-deer is larger than our deer, and the white bear larger than our bear; and, consequently, we must conclude that there are more salubrious climates and better countries within, than any we have yet discovered without.

Hearne also informs us that swans, geese, brants, ducks, and other wild water-fowl, are so numerous about Hudson's Bay, in the spring and summer, that the company every season salt up vast quantities of them, sometimes sixty or seventy hogs-heads.[8] He enumerates ten different species of geese, several of which, (particularly the snow geese, the blue geese, brent geese, and horned wavey,) lay their eggs and raise their young in some country unknown, even to the Indians;[9] as their eggs and young are never seen by them, neither have the most accurate observers been able to discover where they make their winter residence; as it is well known that they do not migrate to the southward; but few of them ever pass to the south, and some of the species are said never to have been seen south of latitude fifty-nine degrees.[10] Most of those fowls molt or shed their feathers in a peculiar manner, in summer, and become nearly naked. Hence it would seem that they must breed in winter while absent, for it is impossible that they could lay and sit whilst molting; whereas, the migratory geese and ducks of this country are not known to shed their feathers, in any great degree; and are well known to raise their young in the summer, whilst in the north. It may, therefore, be inferred, that many of those water-fowls, which Hearne describes, raise their young beyond the icy circle and within the sphere. As many of the ten species of geese he saw there, are unknown further south, it establishes the fact, that they do not come to the south to winter.

In the papers of the Honourable D. Barrington, and Colonel Beaufoy, on the possibility of approaching the north pole, read before the Royal Society of London, there is an extensive collection of instances cited, where navigators have reached high northern latitudes; from which it appears to be well authenticated, that navigators have in numerous instances reached the latitude of eighty-two, eighty-three, and eighty-four degrees:[11] and some are said to have sailed as far north as eighty-eight and eighty-nine degrees.[12] It is almost uniformly stated, that in those high latitudes, the sea is clear of ice, or nearly so, and the weather moderate.[13] To cite the various instances in which navigators have sailed far north, would be too tedious:[14] the whole book principally consists of a series of facts, which have a strong bearing on the subject, and to which I would refer the reader who feels disposed to investigate. The whole appears to strengthen the opinion, that there is a barrier, or circle of ice, about where the whalers go to fish; but, when that is passed, we come to an open sea, and a more temperate region.

The sea is stated to be open, and always clear of ice, even in the middle of winter, on the northern part of Spitzbergen, which is situated in latitude eighty degrees north; and the further north the more clear it is of ice.[15] But, at the same season, on the southern parts of Spitzbergen, the sea is bound up with solid and compact ice.

If the doctrine be true, that the earth is a solid spheroid, the cold must increase regularly as we approach the pole, and, consequently, vegetation invariably diminish: this, however, is ascertained not to be the fact. Nova-Zembla, which is situated in north latitude seventy-six degrees, produces no timber, nor even a blade of grass,[16] consequently, all the quadrupeds which frequent it, are foxes and bears; both carniverous animals. On the coast of Greenland, about latitude sixty-five and seventy degrees, neither timber nor grass grows;[17] while on the northern parts of Spitzbergen, they have reindeer, which are often exceedingly fat; and Mr. Grey mentions three or four species of plants which grow and flower there, during the summer.[18]

On any meridian passing through England, it is ascertained to be more temperate at the latitude of eighty degrees north, than at seventy-three degrees:[19] and both Pinkerton and Barrington inform us, that beyond the latitude of seventy-five degrees, the north winds are frequently warm in winter;[20] that in the middle of winter for several weeks, there falls almost continued rain; and that vegetables and animals are more abundant at the latitude of eighty degrees than at seventy-six degrees.

It has long been observed that the climates vary very considerably on the same parallels of latitude. New York, which is situated in latitude 40 degrees, is known to be considerably colder in the winter than London, which is situated in latitude fifty-five degrees; and the parallel of latitude forty degrees on the plains of Missouri is much colder than the city of New-York. The climate at St. Peters, on the Mississippi, which is in latitude forty-six degrees, is said to be considerably colder than Quebec.[21] This difference of climate has, by some, been attempted to be accounted for, on the principle that land is colder than water, and that the cold is occasioned by the large portion of land in the continent of America: however, I submit to the consideration of the reader, whether so great a difference could arise from a cause of this nature.

In the northern sea, between Spitzbergen and the continent of America, there is a strong current, which always comes from the north, and sets southwardly.[22] It has been stated by some, that, in the spring season of the year, the water of this current is warmer and fresher than the surrounding water of the sea. Various other currents have, at different times, been observed, in different parts of the sea, setting from the north. Floating southwardly on these currents, have been seen large masses of ice, from fresh water rivers, with wolves and bears occasionally on them. New fallen trees have also been seen floating from the north; and various kinds of timber, some of which the species have hitherto been unknown, are frequently found lodged on the northern part of the coast of Norway, having drifted from some region still farther north. Trees have also been found floating in the ocean at latitude eighty degrees; when no timber is known to grow north of latitude seventy degrees. Also, seeds unknown to our botanists, and those of tropical plants have been found drifted on the coast of Norway, and parts adjacent, many of which were in so fresh a state as to vegetate and grow;[23] when it is well known that no plant of their species comes to perfection in any known climate far without the tropics. And, what makes the matter particularly extraordinary, is, that these things appear to be drifted by currents coming from the north; when, according to the old theory, we must believe the sea to be always frozen at the poles, which would render it difficult, if not impossible, to account for the existence of the currents at all.


In the United States of America, and in Europe, the Aurora Borealis is always seen to the north: But many of those travellers and navigators, who penetrated to high northern latitudes, observed the Aurora Borealis in the south, and never in the north. The region in which it is believed to exist, is supposed to be about the place where the verge commences, and about fifty or sixty miles above the plane of the earth's surface; and that the travellers who discovered these appearances south of them, were at that time beyond the verge.

The Indians discovered by Captain Ross, on the coast of Baffin's bay, in the summer of 1818, in latitude seventy-five degrees fifty-five minutes north, when interrogated from whence they came, pointed to the north, where, according to their account, there were "plenty of people;"[24] that it was a warmer country; and that there was much water there. And when Captain Ross informed them that he came from the contrary direction, pointing to the south, they replied, "that could not be, because there was nothing but ice in that direction:"[25] Consequently these people must live in a country not composed of ice; for it appears they deem such an one uninhabitable. Hence we must infer, if the relation given by Captain Ross be correct, that, north of where they then were, the climate becomes more mild, and is habitable; a change, the cause of which is not easily accounted for on the old philosophic principles.

In high northern latitudes, owing to refraction, or some other peculiar circumstance, which hitherto has not, to my knowledge, been attempted to be accounted for, the extent of vision appears to be greatly increased; so that objects, much further than the ordinary distance, are distinctly seen; frequently appearing elevated above the sea, or their real situation; and their image sometimes pictured in the sky. The real objects, themselves, are sometimes seen with the naked eye one hundred and forty or one hundred and fifty miles,[26] and sometimes at the astonishing distance of two hundred miles. These facts are well attested by Captain Ross and other navigators. How this can be accounted for, on the formation maintained by the old theory, I cannot conjecture. I believe it is admitted that the deck of a vessel at sea, any where between the equator and latitude fifty or sixty degrees, cannot be discovered, even by the best telescope, at a greater distance than twelve or fifteen miles.[27] Nay, were there no end to vision, and could the eye penetrate two hundred miles through our atmosphere with sufficient clearness, it would require an observer to be elevated about five miles, before he could discover an object on the surface of the earth two hundred miles distant. But, on the edge of the verge of the polar opening, if the atmosphere were clear, and the power of vision strong enough, an observer might discover objects situated on the verge at any point all round the sphere; as they would be on an exact plane with the observer. And on the contrary, travelling across the verge from the convexity to the concavity of the sphere, a very few miles make objects disappear.

All northern navigators and travellers agree, that high north the sun becomes less bright, and the sky darker, than in more southern latitudes. Is not this owing to the rays of the sun being refracted round the verge of the polar opening? Another circumstance, observed by navigators, who have visited high latitudes is, that the latitude and longitude, as found by celestial observation, frequently differ very materially, sometimes as much as one half, from that given by the log-line.[28] It has also been observed that the mercury in the barometer is less fluctuating in northern regions, than it is further south.

Those appearances observed in the southern hemisphere, which are termed Magellanic clouds, by navigators, have not, so far as I know, been accounted for. They are three in number, of an irregular shape, and observed by night in the South Atlantic, and the south-east parts of the Pacific oceans, (reversed from New-Holland and New-Zealand,) but never visible in the eastern parts of the Indian ocean: their colour is like that of far distant mountains, on which the sun is shining. In the one sea they appear due south, and in the other to the left. They are stationary, appearing perpetually fixed at a certain height, and in a particular situation, as viewed from any given place. The stars and the heavens, in their diurnal revolutions, sweep by them, and they remain the same. To the navigator, who proceeds to the east or west, they appear to be more or less to the right or left of the meridian, in proportion as he changes his longitude; and as he sails south, they increase in height, until they reach the zenith, and finally become north, when seen by an observer south of the straits of Magellan, which is in latitude fifty-two degrees south. Captain Symmes accounts for the appearance of these clouds by the great refractive power of the atmosphere about the polar openings; causing the opposite side of the verge to appear pictured in the sky, as navigators inform us objects do sometimes appear, in the arctic regions; and in the manner Scoresby's ship appeared in the sky, with every particular about her so accurately represented, as to be at once identified by the observers, though the vessel, at that time, was at such a distance as to render it rather incredible how she could be seen at all. As proof of this position, Captain Symmes alleges, that the relative position, shape, and proportions of these clouds, agree in their general outlines with the southern part of New-Zealand, the southeast part of New-Holland, and the whole of Van-Dieman's land, which are situated on, and near to the verge of the sphere, opposite to where the clouds are visible. These clouds are only seen in the night when the atmosphere is clear, at which time the sun is shining on the islands in question. Hence it is alleged, that from these facts, their relative appearance is deducible. As we are never sensible that the rays of light are refracted by the medium through which they pass before they reach our visual organs; we frequently imagine objects to be situated where they really are not; and such is believed to be the case as respects Van-Dieman's and the circumjacent land, as before described.

Franklin, in his journey far north, on the continent of America, discovered a cloud, which appeared to remain always in the same position, and which the Indians informed him was permanent. Not having the book at hand, I cannot now advert particularly to what he says on the subject: but, from memory only, recollect that he states something to that effect. If such an appearance exist there, may it not be accounted for in the same manner as the Magellanic clouds?


Navigators, who have sailed far north, admit the variation of the needle to be excessive. Captain Ross found it in Baffin's Bay, to be as much as one hundred and ten degrees; and Parry, during his voyage in 1822, found it so changed, that the needle pointed within about fourteen degrees of south. All, I believe, concur, that this is a phenomenon which universally occurs in high northern latitudes; but it has hitherto remained unexplained. I believe, according to the old theory, the needle is imagined to be attracted by something at or near the pole: were this supposition correct, the needle would uniformly maintain its polarity on proceeding north, on any given meridian, until you arrived at the very pole itself. The possibility of a moving magnetic cause is difficult, if not impossible, to be reconciled with a solid globe; yet that the magnetic needle does vary on the same meridian, and to a most extraordinary degree, in high northern latitudes, is confirmed beyond all doubt. Why not then urge the variableness of the magnetic cause against the possibility of a solid globe?

According to the doctrine of hollow spheres, this whole mystery, of the variation of the compass, can be satisfactorily explained. The magnetic needle, it is believed, regards the centre of the polar opening, and not the pole or axis of the earth. It will be recollected, that the axis of the earth, being at an angle of twelve or fifteen degrees from the plane of the polar openings, causes one part of the verge to extend farther north than the other, the highest part of which is nearly on a meridian running through Spitzbergen, in about latitude sixty-eight degrees, and the lowermost side in about the fiftieth degree. Now in proceeding north on the first meridian, running near Spitzbergen, there ought to be no variation of the needle until you arrive at the apparent verge, when the needle would cease to traverse; and by proceeding onwards, would turn and point south. Should you proceed north, on a meridian west of this, when you approached the apparent verge, the needle would seem to turn west, but in reality, it would be the meridian turning to the right along the verge to its highest or most northerly point; the needle keeping at a right angle with the verge. And, in like manner, pursuing a course north, on a meridian east of Spitzbergen, on your approach to the apparent verge, the needle would still direct its course at a right angle into the polar opening, (governed, most probably, by some principle of electricity, or other property contained in matter, and kept in one position, subject to the shape of the earth, which may not even yet be exactly known,) the meridian would here wind to the left, and conduct you to the highest point of the apparent verge, north of Spitzbergen. Hence the variation of the needle would be east in Asia, and west in America, which I am told is the fact. From an examination of the variation of the compass, as ascertained in different degrees of latitude and longitude, it increases as you proceed north, and west; which would be exactly the case in accordance with the theory of concentric spheres.[29]

Admitting the earth to be a solid globe, and the cause of magnetism to be some attractive power at the pole, how could the needle vary differently on the same meridian, in different latitudes, at the same period of time, or vary at the same place, at different periods of time? But, admit the doctrine contended for, by the advocates of concentric spheres, and it can be satisfactorily explained. The observations of modern astronomers, have ascertained, that the poles, or axis of the earth, are not always directed to the same fixed star; and, of consequence, that the axis does not always remain parallel to itself. This variation is discovered to be about fifty-one minutes annually; which would make a degree in about seventy-one years: hence the needle always pointing to the polar opening, would vary in about that proportion, at the same place, in the same period of time.[30]

# CHAPTER VI.

*Facts tending to illustrate and prove the existence of a* mid-plane-space, *situated between the concave and convex surfaces of the sphere.*

According to Symmes's Theory, each sphere has an intermediate cavity, or *mid-plane-space*, of considerable extent, situated between the convex and concave surfaces of the sphere, filled with a very light and elastic fluid, rarified in proportion to the gravity, or condensing power of the exposed surfaces of the respective spheres: and also, various other less cavities or spaces between the larger or principal one, and the outer and inner surfaces of the spheres, each filled with a similar fluid or gas, most probably partaking much of the nature of hydrogen. This fluid is lighter than that in which the sphere floats; and has a tendency to poise it in universal space. The spheres, in many parts of the unfathomable ocean, is believed to be water quite through from the concave or convex surfaces to the great mid-plane-space, and probably the earthy or solid matter of the sphere, may in many places extend quite through from one surface to the other, tending, like ribs or braces, to support the sphere in its proper form. Such a formation of spheres appears to be supported by various facts and phenomena; amongst the most prominent of which are Volcanoes and Earthquakes. Many volcanic mountains burst out and burn for ages, discharging from the bowels of the earth immense quantities of lava, pumice, and various substances of various kinds. Some of these mountains have been burning for thousands of years, at least as far back as the records of history have been made known to us.

Had the earth, at its formation, been a solid globe, four times as hard as hammered iron at the centre, and gradually lessening in density towards the surface, we must admit that it would still be solid matter. Governing ourselves by these principles, how can we imagine that such immense caverns, filled with combustible matter, as would be necessary to supply those volcanoes from time immemorial, could have existed? However, that they do exist is certain, which I think is in no way more easily accounted for, than on the plan of a *mid-plane-space*, or of spaces, filled with a certain hydrogenous gas, which being much lighter than atmospheric air, if there should be any small aperture or crevice extending from the surface to the space beneath, the gravity of the outer part of the sphere pressing on it would occasion a portion of this gas to escape through the aperture; and as it comes in contact with the oxygen of the atmosphere would take fire and occasion those tremendous explosions which we know do sometimes take place and cause those mountains to burn for years, until the cavity which supplied the volcanic matter, becomes exhausted; or until some shock or convulsion consequent on the burning, may have loosened rocks or earth of the denser part of the sphere, which falling into the aperture, choke it up. Hence the gas ceasing to escape, the volcano would cease to burn, until some shock or accident should again open the aperture.

The elastic fluid, with which the *mid-plane* cavities are filled, being forced out into the common atmosphere, the greater degree of gravity would condense and set free its latent heat or caloric, and be resolved into its original base, somewhat as coal-gas, out of the tube of a gas-light apparatus, yields up its latent heat by condensation. Hence steam burns when mixed with coal-gas.

If the earth be a solid globe, I am at a loss to account for the principles on which earthquakes occur. Long before I heard of Symmes's theory, or perhaps before it had an existence in the mind of man, when reading accounts of earthquakes, it appeared to me altogether unaccountable, that such violent concussions could take place in one part of the world, and not be felt throughout the globe. It appears altogether inconsistent, that one part of a solid piece of matter, would be shaken so violently, without affecting the whole mass. We are informed by authentic history, that whole islands, and vast sections of country, have been sunk by earthquakes, and never more heard of. On the other hand, islands which are now inhabited, and productive, have been raised, apparently, from the bottom of the unfathomable ocean. How such things occur, I am unable to divine. If the globe be solid, on what principle could a large portion of its surface, which is said to be lighter than the parts beneath, sink into a dense medium? How could a heavy mass, lying a thousand fathoms deep at the bottom of the ocean, rise, and be suddenly elevated above the surface of the water, when all below is so compact, and governed by an opposite and immutable tendency? It appears to be a solecism in nature.

The writer had once an opportunity of witnessing some of the effects of earthquakes. It was his fortune to be on the Mississippi river in the year 1812, at the time when that country was so violently convulsed with an earthquake. He saw and heard innumerable explosions, as though a large quantity of air had been confined in the bowels of the earth, and, seeking vent, rushed out with a tremendous sound; forcing up considerable quantities of sand through the apertures, in many instances mixed with black muddy water, and a substance resembling stone coal, or carbonated wood, which emitted a strong bituminous odour, when exposed to fire.

At one place the river was stopped in its course a short time: the water rose to a considerable height above its common level; and, on the west side of the channel of the river, there was a counter-current for a few minutes of an astonishing velocity. So great was its force, that for some distance the cotton wood and willows on the margin of the river, were either prostrated or bent up the stream; and their branches looked as if they had been dragged a long way on the ground. The waters of the river soon subsided, and flowed in their natural direction.

So tremendous were those explosions, that when happening under large trees, the tenacity of their texture yielded at once to their force; and the largest in the forest were split and fractured from root to top. During these convulsions, the ground on which the town of New-Madrid is situated, together with the country for several miles round, sunk about five feet below its former elevation; in which situation it has remained. Eight years afterwards the writer was again on the same spot. The desolate aspect, which the country presented at the time he witnessed those scenes, was measurably obliterated: but the banks of the river were still in their sunken situation.

How could all those violent convulsions take place at this point, and not be felt at New-Orleans, along the sea coast of the United States, and other places? Whence came this water and air, which issued from those apertures in the earth? And why did the river for a few minutes flow in a contrary direction, and then resume its natural course? If the earth be a compact and solid globe, I can account for none of these things; but admitting the formation of the sphere to be such as I contend for, they are all resolved into the most simple principles; and what would otherwise be impenetrable mystery, is made as plain as noon-day. If the sphere be formed as I allege, those concussions were doubtless occasioned by the gas or fluid in the *mid-plane* or some intermediate space, near the surface, which, by being suddenly rarified, would make it expand, and cause the upper part of the sphere to be suddenly elevated in the neighbourhood of the Little Prairie; and hence the waters of the river, pursuing the laws of gravity, would flow in a contrary direction. This sudden expansion, and elevation of the surface, would cause apertures, through which the rare gas would escape, and the surface would then settle down again, not only to its former level, but, as a considerable portion of this gas had escaped, the remaining part would occupy less space; hence the surface of the country, around New-Madrid, would be below its former situation.[31]

The fluid, or gas, which fills the mid-plane and intermediate cavities, is most probably the same, or partaking of the same nature, (though perhaps in a purer state,) with that which oozes out of fissures in the earth, at the bottom of deep mines, called by chemists, *hydro-carbonate;* which being highly inflammable, takes fire from the lamps used by workmen, and explodes with such violence as to destroy both men and horses employed in the mine. This is a frequent occurrence in the deep coal mines of England; and great numbers annually have lost their lives in this way, before the introduction of Sir Humphrey Davy's lamp. I am also informed, from good authority, that the miners, in some of the deep coal mines in England, once felt, or heard an earthquake, which happened in Italy, whilst those on the surface of the ground had no knowledge of it. This would be the case, if the intermediate cavity, which caused the earthquake, extended in that direction, and near the bottom of the mine; as it is presumed the rare gas with which those spaces are filled, is better adapted to the conveyance of sound, or vibratory motion, than the more solid parts of the sphere, or even the atmosphere around us.

On the supposition that the globe is solid, and the matter composing it at rest, as respects itself, on what principle can boiling and hot springs be accounted for; some of which issue out several thousands of miles distant from where any volcano or subterranean fire is known to exist; particularly as to those on the waters of Red river, in the state of Louisiana, which are sufficiently hot to cook meat in a few minutes.

Phenomena which occur in various lakes in Europe, may be adverted to in support of this theory. The waters of lake Zirchnitzer, in the Dutchy of Carniola, in Germany, flow off, and leave the basin empty; and again fill it, in an extraordinary and impetuous manner; bringing up with its waters fish and even sometimes wild water fowl.[32] In the same country, there is a subterranean lake, in the Grotto Podspetschio, of considerable extent; the whole of this vast body of water, at certain times, will disappear in a few minutes, and leave the basin dry; and after a few weeks, it again suddenly returns, with a frightful noise. The lake of Geneva, and some others in Switzerland, at certain times rise and fall several feet without any cause, which has as yet been satisfactorily explained; and some writers inform us, that those lakes, particularly Geneva, send forth, at times, a grumbling noise. In the Saian mountains, near the source of the Yenisei, is a lake, called Boulamy-Koul, which, at the approach of winter, emits strange sounds, somewhat similar to those which precede the eruption of a volcano, and which are compared by the neighbouring inhabitants to howling. The inhabitants on the borders of Baikal, also state, that they have often heard dreadful and terrific howlings proceed from that lake.[33] The lake, Agnano, in Italy, sometimes, especially when the waters are high, appears to boil at its borders. This ebullition is supposed to be occasioned by some gaseous fluids, discharged into the bottom, which traverse the waters of the lake.[34] These various phenomena, which cannot be easily accounted for, might be best explained perhaps, on the principles of *mid-plane-spaces*. In various parts of the north, thick strata of ice are found, under a thick soil; and on ice-bergs, floating in the ocean, have been discovered masses of earth, of granite, and of other rocks.[35]

On the shores of Greenland the ebb tide flows towards the coast, apparently as though it passes under the land, and the flood tide recedes from the shore; and in those regions the sea is almost universally found deeper as you approach the shore.[36] When the whales become scarce, experience has taught the whalers to seek for them near the shore, as if at certain seasons they retired to it, and then disappeared. Captain Symmes imagines that the sea extends quite through the spheres, about Greenland, and that the whales suddenly migrate either to the *mid-plane-space*, or to the seas on the opposite side; which he alleges to be the case with several other species of fish, as well as seals; all of which, he supposes, breed in the *mid-plane-space*. The reasons that induce him to adopt this conclusion are various; such as, that fish have been thrown up by the eruption of a volcano in South America[37]—herring appearing in such immense numbers at certain seasons of the year—the whales seeming to pass under Greenland—two seals having been once caught in Lake Ontario, which is said to be unfathomable, although this lake is many degrees south of where the seals have ever before been known to come—and the various species of fish in our northern lakes which appear and disappear at certain periods. That the exterior seas in some places communicate with the interior seas, is rendered probable by various other circumces; such as currents running continually into the Mediterranean, and no visible outlet to the water thus continually flowing in. It is scarcely probable that evaporation could carry off all the water supplied by the straits of Gibraltar—the white sea being more salt at the head than at the foot—the tides being higher in the Baltic than the Mediterranean—white foxes having been forced up by the waters of the sea (as Symmes undertakes to prove) in the northern regions—the peculiarities of the tremendous whirlpool on the coast of Norway, called the Maalstroom, which sucks in, and discharges the waters of the sea with great violence—and those observable in the Bay of Biscay, which are said to be unfathomable.

# CHAPTER VII.

*Several objections, made to the Theory of Concentric Spheres, answered, particularly the one that it contravenes religious opinions; demonstrating that the earth, and the other orbs of the universe, are formed on the best possible plan for the maintenance and support of organic life.*

Some of the most prominent objections which I have heard advanced against the theory of concentric spheres are the following:

1st. That if the earth be not a solid globe, but a hollow concentric sphere, the quantity of matter being diminished, the attraction of gravitation must be lessened so much that all moveable bodies resting on the earth would be thrown off by centrifugal force, in the line of a tangent from the surface of the sphere.

2d. That according to the established laws of gravity, a hollow sphere could not exist in nature: that matter would be gravitated to the centre, and particularly about the polar openings, so as to make it collapse.

3d. That if the orbs were hollow spheres, the mutual influence of the planets on each other would be so far destroyed, that they would cease to revolve in regular orbits.

4th. That the interior of the sphere can never receive the light and heat of the sun; is involved in perpetual darkness, and more suited to the infliction of punishment on perverse and rebellious spirits, than for the residence of beings, fitted and designed for the pursuit and enjoyment of happiness.

5th. And finally, the adherents of the new theory have been charged with atheism, deism, and such like epithets, as though they intended to overturn the works of God, and thwart the laws of nature.

1st. As to the first objection, I would enquire, has it yet been ascertained with mathematical certainty, in what exact proportion one particle of matter attracts another? And may there not be some law of nature with which we are not yet well acquainted? All the experiments, hitherto made on the attractive power of gravity, were made on the principle, and under the belief, that the earth is a solid globe; and consequently the deductions were drawn accordingly. Suppose the attraction of gravitation, inherent in matter, to be so much increased, that a hollow sphere would possess the same attractive power, as if it were a solid globe, would not all the results and consequences be exactly the same? This being the case,—and I know no reason why we should conclude differently,—the whole force of the objection appears to fall to the ground. According to Newton's principle of gravity, the matter of the sphere would attract all particles of matter placed on the surface, as well upon the concave as convex, in nearly equal proportions; and the centrifugal force, which, on the outer side of the sphere, tends to throw bodies off, on the concave side, would have an opposite effect. Hence, a person standing, or trees growing, on the interior surface, would be in no more danger of being precipitated to the next sphere, between them and the centre, than those on the outer part of the sphere, when they should be *turned* (what is familiarly called) *down*.

The experiments made on the density of the globe, by observations with the plum-line, at the foot of a mountain, are very ingenious; but they must be subject to great uncertainty. The true deviation of the plum-line, the exact quantity of matter in the mountain, or, indeed, the quantity of matter between the plummet and the centre of gravity, are points difficult, if not impossible, to be ascertained with mathematical precision.

If the attraction of the sun is just sufficient to keep the earth in its orbit, what can give the tendency to retain Jupiter and Saturn in theirs, each of which, if solid, contains such a vast quantity more than the earth, and removed to so great a distance from the sun, that his influence upon either must be greatly lessened by both?

2d. As to the objections that a hollow sphere of the dimension of the earth cannot exist in nature, I can discover no sound reason to warrant such a conclusion. Many hollow cylinders and spherical figures, we know do exist on the surface of the earth; and notwithstanding their own gravity, which the different parts exert on each other, as well as the gravity of the earth, they retain their shape and position; and had the matter in the earth originally been thrown by a centrifugal force into the form of a hollow sphere, or had the first creating power originally given it that shape,—I can discover no good reason for a change; neither should I entertain any apprehensions of the particles of matter coalescing at the centre.

3d. The force of this objection I cannot appreciate; for if all the planetary orbs in the universe are composed of hollow concentric spheres, they must exert the same relative influence on each other, which they would if they were solid orbs, as they would each contain the same proportion of matter as respects each other. Hence no good reason appears why a system of hollow concentric spheres might not do just as well, and perform their revolutions with the same regularity, as a system of solid ones.

4th. This great and alarming objection comes next:—that we are about placing a world in eternal darkness, cut off from all the comforts and pleasures of refined life, for the enjoyment of which we are so eminently qualified. Let us examine the force of this objection; and if we cannot show that the interior is, at least in some degree, illuminated, we must then conclude that it is a very dreary abode, and unfitted for the residence of beings so fond of light as we *profess to be*.

According to the new theory, the northern polar opening is about four thousand one hundred and fifty miles in diameter, and the axis of the earth is at an angle of about twelve degrees with the axis of the plane of the polar opening; consequently, as the sphere revolves on its axis, one side of the verge of the polar opening will extend considerably further north than the other. The verge of the north polar opening on the low side, is laid down at about fifty degrees of latitude, and the verge of the high side at about sixty-eight degrees.

Now, supposing the sun to be exactly of the same diameter as the earth, and placed directly over the equator, when the low side of the verge was turned towards the sun, the direct rays from his northern limb, independent of refraction, would pass the edge of the lower part of the verge, and fall on the inner part of the sphere, on the concave part of the high side opposite, as far as eighteen degrees, or upwards. When the sun would be on the tropic of Cancer, in June, he must then throw the rays from his centre twenty-three and a half degrees further within the sphere, or within twenty-six and a half degrees of the equator; but the diameter of the sun being so much greater than the earth, the rays from his northern limb, would fall about thirty-three minutes further within the sphere, and leave not quite twenty-six degrees between that and the equator to be excluded from his direct rays. This relates to the northern polar opening; as to that of the south, which is believed to be much larger, we will make a few remarks. The lower side of the south polar opening, is laid in about latitude thirty-four degrees, and the higher side, in about latitude forty-six degrees. Were the sun of the same diameter with the earth, as above premised, and placed on the equator, his direct rays would be thrown into the south polar opening when the low side was towards him, about twelve degrees, or to within thirty-four degrees of the equator, and when on the tropic of Capricorn, in December, twenty-three and a half degrees further, that is, the inner part of the southern hemisphere of the sphere, on the high side, would be lighted thirty-five and a half degrees within the verge; and the direct rays of the sun would shine within ten and a half degrees of the inner centre of the sphere or equator. These observations, you will observe, are made in the most unfavourable point of view. It is well known, that the diameter of the sun, is vastly greater than that of the earth; consequently, his rays would pass into the polar opening so much further, in proportion as the angle of his diameter, and that of the earth, differ, which would be about thirty-three minutes further, bringing his direct rays in the south, within less than ten degrees of the equator; and this would be the case as the sphere revolved on its axis, once in every twenty-four hours. When the sphere turned, with its high side towards the sun, it would be night, or twilight, and when the low side was next the sun, it would be day; at all events, the direct rays of the sun would fall on a space of about thirty-six and a half degrees in breadth; the reflection from which would light the whole of the remaining portion of the inner part of the sphere, to a greater degree, than any moon-light with which we are acquainted. But there is another circumstance which tends to throw the rays of the sun much further into the concave than we have yet got them; that is, the refractive power of the atmosphere. It is a well known fact that the rays of light are very much refracted when passing out of a rare into a denser medium; and about the poles of the earth it is believed, (and this belief is confirmed by navigators) that refraction increases very considerably, owing to the great density of the atmosphere. We have good reason then to believe that refraction throws the rays of the sun several degrees further within the sphere. But let us take the known refraction of the horizontal ray, at or near the equator (say one half of a degree) it would throw the rays of light so much further into the concave, and not leave quite thirty-seven degrees in the centre of the sphere deprived of the sun's rays. The motion of the earth causes the apparent motion of the sun to be about fifteen degrees in an hour, as the diurnal revolution of the earth causes the sun to move apparently through three hundred and sixty degrees in twenty-four hours. Now it is a well known fact to all that the sun gives us light sufficient to be called day-light, for about an hour after he descends below the horizon; consequently he must afford us light when he is fifteen degrees obscured from our view. Accordingly, the sun, though he might not be visible, would illuminate the concave part of the sphere fifteen degrees further than his direct rays fall, which reduces the space in the interior of the sphere to the breadth of not quite seven degrees which would still remain unlighted.

But this is making calculations on the most unfavorable premises possible. Considering the form of the earth, and the power of refraction, I have no doubt but the direct rays of the sun would fall on every part of the inner sphere. However, I have proceeded on such premises as, I conclude, the most sceptical must admit. Light, we know, is reflected from solid bodies on which it falls, and also from the atmosphere: the rays of the sun, then, which would pass the lower part of the verge and fall on the opposite concave surface, would be reflected back in all directions, and most probably light the whole of the interior of the sphere sufficient for the ordinary purposes of life. By way of further illustration, suppose a perpendicular wall were raised on a plain, one mile high, does any person believe that there would be no light on the side of the wall opposite to the sun; although his rays would have to form an angle of one hundred and forty, or one hundred and fifty degrees, to reach the earth on that side of the wall? No axiom is more evident than that the rays of light are communicated to other places than those on which the rays of the sun fall directly; for example, we all know that a close room, however large, with a north window, will be sufficiently lighted by refraction and reflection from the atmosphere, provided there is no obstruction opposite the window, although the rays of the sun would have to form an angle of one hundred and fifty degrees to enter it, and why might not the whole interior of the sphere be lighted in the same manner, even supposing the rays of the sun should never enter directly. The north polar opening being about four thousand one hundred and fifty miles in diameter, and the southern six thousand three hundred and fifty, with the whole force of the direct rays of the sun falling on and passing through the atmosphere at either polar opening, it would not require refraction, or reflection, to make an angle of ninety degrees to light the whole of the interior concave; and certainly the polar openings are sufficiently large for the purpose, when we compare a common window with the dimensions of an ordinary sized room.

It is believed, by the adherents of the new theory, that the atmosphere, within the concave, and about the polar openings, is much denser than our atmosphere; which appears inevitably to be the case, as the centrifugal force on the convex has the tendency to throw the atmosphere *from* the surface, and on the concave to force it from the centre of motion, and nearer *to* the surface. This admitted, the rays of the sun passing out of a rare medium into a denser, would be refracted much further into the sphere; and the sun-shine on the surface of one sphere would be reflected obliquely, according to the angle of incidence, to the next sphere, and in this manner might be extended even beyond the centre of the concave. It is also believed, that near the verges of the polar openings, and perhaps in many other parts of the unfathomable ocean, the spheres are water quite through, (at least all except the *mid-plane-spaces*, or cavities) which being the case, light would probably be transmitted between the spheres.

The apparent elevation of celestial bodies above their true altitude, is greatest when the body is on the horizon, which is ascertained to be a little more than half a degree; hence, in our climate, the sun appears three minutes sooner, and sets three minutes later than is really the case, which increases the length of our day six minutes, by refraction. This gradually increases in proceeding from the equator to the frigid zones; and at the poles, were the earth entire, the day should become thirty-six hours longer, by refraction alone, than it would otherwise be.[38] It was doubtless owing to some peculiar refractive power in the northern regions, that caused the Dutch, who wintered on Nova-Zembla, (which is in latitude between seventy and seventy-eight degrees,) on the approach of summer, to see the sun about two weeks sooner than he should have appeared in that latitude, according to astronomical calculation.[39] This tends to show that there is more refraction in the northern regions than is observable in the south.[40]

From an attentive examination of these considerations, I am induced to conclude, that the interior of the sphere may be as well lighted as the exterior; or at all events, if not favoured with so great a degree of light at all times, it has a more regular and constant supply. But, admitting every thing on this subject that the opponents of the theory can suggest, I still discover no substantial reason why the earth may not be a hollow sphere. I can see no substantial reason why the inhabitants of that portion of the earth, (if any exist there) should be furnished with as great a degree of light, and as intense a heat, as we have upon the convex part of the sphere. Must it of necessity follow, that it cannot be inhabited, or if inhabited, that the beings who people its surface, are less happy than we? Certainly not. Is it not well known to us, that every grade and species of animals, under every variety of circumstance, whether inhabiting the air, the earth, or the water, are fitted by an all-wise Providence to their several conditions, and mediums, in which they reside? As well might we conclude, that the immense planet Jupiter, situated so far from the sun as he is, can be nothing but a dark, cold, and barren waste, unfitted for the residence of intelligent beings. It is ascertained by calculation, that the light and heat which Jupiter receives from the sun, is only the one twenty-seventh part of what the earth receives.[41] The light and heat which Saturn receives from the sun is estimated at only the one hundredth part of that of the earth;[42] and the planet Georgium Sidus, revolving such an immense distance further from the sun, than either of them, must enjoy still less light and heat; according to which, we would conclude, (if we adopt the belief, that the degree of light and heat, to which we are accustomed, is necessary for the support of life,) that those vast planets are not fitted by the God of nature for the residence of intelligent beings; however, I am inclined to believe that both light and heat are communicated to them, in some way not well known to us, sufficient for the purpose. The true principles of light and heat, and the manner in which they are generated and transmitted, are not perhaps yet well understood and defined.[43]

5th. Others, when the new theory is mentioned, cry Atheist, Deist, blasphemy! as if its advocates proposed to make a new world, and support it without the intervention of Divine Providence: such opponents scarcely deserve an answer. It is believed by all, that the earth, the sun, the moon and stars, are the work of an Almighty power. Whether solid globes or hollow spheres, they equally owe their existence to the great first cause, that spoke matter into existence, that arranged it in whatever form and order infinite wisdom dictated; and that still supports and governs the whole by universal and unvarying laws. But it is as well known, that the Almighty Disposer, interposes no miracles for the accomplishment of his designs, but makes use of means that are uniform in their application, to effect the intended purpose; hence Geologists, Philosophers, and Astronomers, attempt to account for the existence of all matter, and for the formation of planets, according to what is believed to be the established laws of matter. In so doing, we do not disparage the wisdom of the Creator, nor controvert the truth of that divine record, which Providence, in his goodness, has given us for our rule of life. True it is, the sacred scriptures give us very little information relative to the structure and formation of the earth and the other planets. They were not intended to teach mankind Geology, Geography, or Astronomy; yet where assertions are clearly and distinctly made respecting these things, we have reason to believe them literally correct: as for instance, when the Psalmist informs us, that God hung the earth upon nothing; that He balanced it in empty space, we are to look for corresponding facts; though it was at variance with the opinion of the world at that time, modern astronomy now teaches that such is the fact. In like manner, when we meet with assertions, such as that "the fountains of the great deep were broken up," (והארץ היתה תהו ובהו, chapter 1, verse 2,[44]) we must acknowledge their correctness; and I think it will be admitted, that they are at least as much in favour of this new theory as the old.

The skilful and attentive observer of nature, whether examining the most minute or the most sublime, will discover that infinite wisdom, judgment, and ingenuity, equally prevail throughout. The principal aims of the great author of all things, appear to have been animation, diversity, and usefulness; the air we breathe, the water we drink, the vegetables on which we feed; indeed every leaf and plant of the forest and field—all teem with animal life. Why then should we believe, or even presume to think, that the Almighty Fiat, which spoke matter into existence, for the support and maintenance of living creatures, innumerable, and endless in the variety of their organization, their colours, their passions, and their pursuits—why, I say, should we then presume, that the omnifick word would create even the smallest particle of any of the immense, the innumerable orbs in the universe, of inert or useless matter, devoid of activity and design? This earth, when compared with the magnitude and number of other planets we know, is but as a point; yet we can hardly conceive, small as she appears by comparison, that she was only designed to have animate life on her surface, and all the rest to remain useless! Such an idea seems unworthy of the Divine Being, whose essence is all perfection. Can we for a moment suppose, that the interior parts of the earth, have received less attention from the Creator, than the objects which are under our immediate inspection? On the contrary, may it not be more rationally inferred, that, for the object of more widely disseminating animation, spheres are formed within spheres, concentric with each other, each revolving on its own axis, and thus multiplying the habitable superfices?

Great and sublime as our conceptions of the Deity must be, when we contemplate the earth and its inhabitants—if we turn our attention to the solar system, our world dwindles into a little insignificant ball. Yet if we cast our eyes still beyond, and contemplate the eighty millions of fixed stars, which a good telescope brings to our view, each the centre of a mighty system of revolving worlds; and then reflect that all this is only one little dark corner of creation, we are lost in the magnitude of the contemplation. But when we come to consider each of these fixed stars, with their planets, and they with their satellites, all consisting of concentric spheres, revolving within each other, in due order, and adapted to the support and comforts of life, for countless millions of beings; we are struck with ten-fold astonishment and admiration, and bow with reverential awe, before Him who sits at the head of the universe, and governs the whole by unvarying laws. It would seem to me, that in contemplating this new order of creation, the imagination must break through and soar beyond its old boundaries. It would seem that on embracing this doctrine, the spirit must expand with increased devotion, and be entirely absorbed in the infinite wisdom and power of Him, who was competent to devise, and able to execute, such a beautiful arrangement of matter.

# CHAPTER VIII.

*General observations on the Theory of Concentric Spheres, with a few suggestions to the Congress of the United States, to authorize and fit out an Expedition for the discovery of the Interior Regions; or, at least, to explore the northern parts of the continent of America.*

Of the many various and conflicting theories which have been advanced, relative to the form, structure, and motion of the earth, the theory of Concentric Spheres deserves to rank as one of the most important: for, should it hereafter be found correct, the advantages resulting to the civilized and learned world, must cause it to stand pre-eminent among the improvements in philosophy. The habitable superfices of our sphere would not only be nearly doubled; but the different spheres of which our earth is probably constituted, might increase the habitable surface ten-fold.

That such may be the construction of the earth, every law of matter with which I am acquainted, seems to admit, at least of the possibility; the different appearances of the other planets render it probable; and the various concurring terrestrial facts existing in the arctic regions, to my mind, render such a conclusion almost certain. And further, that matter and space are never uselessly wasted, is an axiom, not only of sound philosophy, but of natural religion, and of common sense.

Many of the theories which have been advanced respecting the earth, are vague and uncertain, and will remain so forever; being predicated on deductions drawn from certain premises that can never be established with certainty; consequently they must rest wholly on the strength of the arguments drawn from the premises, as they are not susceptible of being demonstrated by experiment. Not so with the theory of concentric spheres. Its correctness admits of ocular demonstration. The interior of the sphere is declared accessible, and the whole extent capable of being accurately explored; thereby establishing the theory, or disproving and putting it at rest forever.

The celebrated Dr. Halley, in the year 1692, in his attempt to account for the change of the variation of the magnetic needle, advanced a novel hypothesis, as respects the internal structure of the earth. He supposes that there is an interior globe, separated from the external sphere by a fluid medium; or that there may be several internal spheres, separated from each other by atmospheres, and that the concave arches may in several places shine with a substance similar to that which invests the body of the sun, producing light and heat for the accommodation of those internal regions which he alleges may possibly be inhabited by animate beings.[45]

However, he suggests no idea of Polar Openings, nor of any communication from the outer surface to those interior regions; consequently their existence must have remained forever a matter of mere conjecture.

We find that Dr. Halley, in the wisdom of his philosophy, believed those internal regions to be lighted, though situated many thousand fathoms beneath the surface, and without any aperture to communicate light from without. Why not, then, believe that the interior of the spheres, according to Symmes's theory, may be lighted, when he lays down such vast openings at either pole for that purpose?

Euler was also an advocate for the theory of Dr. Halley. He believed, with him, that the earth is hollow, with a ball, or nucleus, included in the centre; he, however, differed from Halley as to the nature of the nucleus. Halley believed it to be constituted of the same materials of the exterior crust of the earth. Euler believed it to be a luminous body formed of materials similar to the sun, and adapted to the purpose of illuminating and warming the interior surface of the crust, which he supposed might be inhabited equally with the exterior surface. He fancied that this luminous ball had no rotary motion, and that the outer shell revolved around it. However, neither he nor Dr. Halley left any opening by which the internal regions could be explored; their existence was therefore left to rest on vague hypothesis.[46]

These different theories, however extravagant they may appear to us, were believed and supported by those men, whom we must acknowledge were among the most learned of the age in which they lived; and among the mathematicians in Europe they have yet some warm supporters. Why not then give Symmes's theory of open poles, and *concentric spheres*, a serious investigation, the correctness of which is so much more probable, and the demonstration of its truth or falsehood so much more practicable? At all events a *voyage to the polar regions*, with an eye to the accomplishment of Symmes's purpose, might be productive of incalculable advantages to the cause of science in general. With respect to astronomy and geography, it would afford many new lights, and perhaps discover and establish many new principles, not thought of at this day.

"*Knowledge is power*," and so far as an individual acquires a knowledge of literature and science, above his contemporaries, so far does he possess a power and influence over those among whom he resides. So does a nation, when she becomes characterized for the acquisition of knowledge in the sciences and the arts. Those nations which have made great and important advances in the improvement of science, or in new discoveries, have acquired a pre-eminence of character and standing, among other nations of the world.

The United States of America, having assumed a respectable station among the nations, is fast advancing in wealth and power. Her territories are stretched over a vast extent of country; and her population is increasing with a rapidity unprecedented. We are already looked up to, by other nations, as a people of very considerable importance; and as having made a successful experiment in politics and government, which politicians had before considered impracticable. Ought we not then, as a nation, (paying some attention to the progress of science and knowledge,) to hold out inducements for the progressive improvements, and useful discoveries of our own citizens?

While the English, the Russians, and the French, are making great exertions for the purpose of discovery, and the advancement of science; will America remain idle and inactive? Will she adopt the unwise policy that individual enterprise ought to be let alone? Other nations act differently; and they have long been directing their researches towards the acquisition of a more perfect knowledge of our globe: and such exertions have always been considered as the most glorious actions on record in the annals of their history. By so doing, they have not only been amply rewarded themselves, but have benefited the world at large, by the acquisition of important information respecting the before unknown parts of it, and by the improvement of science. Will America then sit by inactive and contented, while she is surrounded with plenty, and enjoying a situation most enviable in the career of nations? Let us rather encourage than shackle the genius and enterprising spirit of our own citizens; and not act like an avaricious miser, who directs all his thoughts to the calculation of dollars and cents. Had this "let alone policy," been pursued by the nations that have sent out ships of discovery, what would have been the situation of the world at the present day? Bounds would have been set to the great field of philosophy, and the arts and sciences must have flourished only within a circumscribed sphere. In vain might the revolving planets have forced upon the minds of mankind their beautiful order, motions and attractions;—the extensive continent of America, must yet have remained a gloomy wilderness; and the wild flowers have bloomed upon her fertile plains, only to be crushed by the foot of the unlettered savage.

If we take a retrospective view of the world, for some centuries back, we shall find the knowledge of the most scientific nations, bounded by a circumference of two or three thousand miles. At length a few enterprising individuals, aided by their governments, made extensive discoveries:—A Columbus discovered the vast continent of America; and subsequent navigators discovered the extensive countries of New-Holland, New-

Zealand, and numerous islands in the Pacific ocean and South sea. All of these now disclose to us, that what was formerly believed to constitute the whole habitable world, is but a spot, one little corner, in the parts known at this day. Even yet, a vast portion of our globe remains unexplored. Why then should we contribute nothing towards the attainment of the grand pursuit of nations? We, who are destined, I hope, one day to stand as the first nation under the sun—Why should we fold our arms and sit inactive, while that little spot Great Britain, is making such efforts to explore those regions?

It would not be an unwise policy, for the American government to foster and encourage such noble workings of genius. It can in no way be inconsistent with the present policy of our government, that an expedition should be fitted out to explore the polar regions; but, on the contrary, it would bespeak a spirit of liberality, and a desire to promote scientific enterprize. It is neither against the constitution nor laws of our country; we are now at peace with the world; taxes are comparatively trifling; the situation of our country at present affords a most favourable opportunity for the accomplishment of the undertaking. It is one of such importance too, as will justify the use of money and men; while the honour of the discovery of a New World would be its reward.

I apprehend that we only lack confidence in our own abilities, to perfect and explain many things not dreamed of by the ancient philosophers. We are inclined rather to undervalue our own efforts; and, like our former opinions on manufacturing subjects, think we can never appear to advantage, unless dressed in a coat of foreign manufacture. It appears to savour of the doctrine, that no new opinion or proposition can merit attention, or be adopted, unless it come from a European source. Had the proposition of concentric spheres, or a hollow globe, been made by an English or French philosopher, instead of a native of the United States, I very much question, whether so large a share of ridicule would have been attached to its author and adherents.

It may be replied, that the idea of a world within a world, is absurd. But, who can assert with confidence, that this idea is, in reality, nothing more than the imagination of a feverish brain? How is it shown that such a form does not exist? Are there not as strong reasons for believing that the earth is constituted of concentric spheres, as the court of Spain, or any man in Europe, had to believe that there was an undiscovered continent? Has not Captain Symmes theoretically proven his assertions of concentric spheres and open poles, and embodied a catalogue of facts, numerous and plausible, in support of his opinions? And who has confuted his assertions? I *dare* to say, that none can be found, who *can* fully disprove them, and account for the facts which he adduces as the proofs of his theory. Is there not the same reason to believe, that the earth is hollow, as there is to place implicit confidence in the opinion, that the planets are inhabited? And yet the one has been ridiculed as the wild speculations of a madman, while the other receives credit among the most enlightened.

If it can be shown that Symmes's Theory is probable, or has the least plausibility attached to it,—nay, that it is even possible,—why not afford him the means of testing its correctness? The bare possibility of such a discovery, ought to be a sufficient stimulus to call forth the patronage of any government. And should the theory prove correct, and the adventure succeed, would it not immortalize our nation? The fame of Symmes, and his native country, would only expire with time! But, even should the expedition fail in the main object, there would still be neither loss nor disgrace. If the interior world have no existence but in Captain Symmes's imagination, would it be a matter worthy of no consideration to explore the northern parts of our own hemisphere? In the attempt, we might discover something of great importance—in chasing a phantom, we might hit on a reality—in searching for the "unknowable," discover what has hitherto been unknown; some new islands; some undiscovered sea; some north-west by west passage, or inlet; some new phenomenon of nature; some hitherto unknown inhabitants of the polar regions; nay, even the pole itself. And would it be a matter of no consequence, that a citizen of our own country should first stand on the axis, and plant the stars and stripes of our own country beneath the polar star? And should this be effected, will not the glory and honour our nation would acquire thereby, be worth the expenditure? No one, I hope, will say that it would not be worth it all, ten times told. But in case *this* should fail, would it be a matter of no consequence, to explore the northern parts of our own continent, and fill up the blank on the map of the northern hemisphere? This, in my humble opinion, is far from being impracticable. A steam vessel might run from the mouth of the Oregon river, and proceed along the north-west coast of America through Behring's Straits, round to the Atlantic; or, if impeded by ice, a party might pursue their journey on foot, with sledges, on the ice, and along the coast quite round to Hudson's Bay. The accomplishment of this, I deem no chimera. The writer of this, for one, (and he has no doubt Captain Symmes, and a sufficient number of others) would volunteer to accomplish the enterprise. And should such an expedition be authorized and fitted out by the government, rest assured, if they did not penetrate the interior of our sphere, or plant the American standard beneath the *great Northern Bear*, they would at least furnish a correct map of the coast of America, from the mouth of Oregon round to fort Churchill;—or make the snows of the north their winding sheets.

Within a few years, several expeditions have been fitted out for the purpose of discovery, by different nations in Europe, and particularly by the English. Ross, and Parry have visited the arctic regions; and Parry now is out on his third voyage, as though there were some hidden mystery there, which the English government is anxious to develope. It is not likely that they would have fitted-out, and dispatched four successive expeditions, merely to view Ice-bergs and Esquimaux Indians. As for the discovery of a north-west passage to the East Indies, it cannot be their sole object, as the continent of America has been explored by land to seventy-two degrees of north latitude; and, according to the old theory, beyond that latitude the seas are so incumbered with ice as to render their navigation extremely difficult, if not impracticable; from which, I am induced to believe, that they have discovered something in those regions which indicates a state of things different from that heretofore believed to exist.

Under the protection of the Russian government, Kotzebue, and Baron Wrangle, have been engaged in similar enterprizes, and although these different attempts have afforded considerable light on the subject, yet they are rather calculated to awaken than satisfy curiosity. Many of the facts, however, which are urged as proof of the theory of concentric spheres, have been confirmed or corroborated by the personal observations of those skilful navigators. But so long as they lack confidence in the theory, it can scarcely be expected they will make the discovery; the winding meridians which they will pursue, when intending to proceed straight forward, will keep them bewildered among the ice, along the circle of the verge, or finally bring them out towards the exterior surface of the sphere, no wiser than when they set out.

As yet, we are more indebted to other nations, than our own, for a knowledge of the continent of America. A knowledge of the north-west coast is interesting to the civilized world at large; but to none more so, than the United States; and I humbly think, that the honor and interest of this confederated Republic, are more deeply involved in this subject of making discoveries in the northern seas, than any other nation's can be.

Should a voyage of discovery be undertaken by our government, it is hoped that the northern coast of the continent of America will, at least, be examined. The undertaking would not only redound to the fame of our country, and to that of the individual entrusted with the enterprise, but must be productive of immense advantage to our commerce and national prosperity; and carry our "star spangled banner" among a people with whom the civilized world, as yet, have had no intercourse.

The prosecution of such an enterprise would be attended with no very considerable demands on the treasury; the employment of one or two of our ships of war, now in commission, for the object, would cause little additional expense. But, even admitting that a few thousands, or even hundreds of thousands, would be necessary; of what importance is it, when weighed against the magnitude of the object to be accomplished? Could our public vessels be better employed, than in surveying our north-west coast, and in discovery? Our naval officers would rejoice on seeing opened to their view a new path to fame, independent of the acquisition to their nautical experience. Many of our brave and skilful navigators would be proud of an appointment in such an enterprise; many naturalists and men of science, would cheerfully, at their own expense, if necessary, accompany such an expedition. And although we may not expect such an enterprise to be accomplished to the full extent of Captain Symmes's anticipations, and those who believe in his doctrines; yet, as Americans, we cannot but wish that the theory, which has been first advanced by a fellow-citizen, should be countenanced by our own government, and tested by the citizens of our own country.

# CHAPTER IX.

*A few brief suggestions, relative to the description, tonnage, and number of vessels, necessary to be equipped for a voyage of discovery to the interior regions of the earth; the number of men necessary to be employed on board, articles necessary for the outfit, and the probable expense attending the same; also, as to the route most proper to be pursued to accomplish the object of the expedition.*

Captain Symmes, in his first circular, published at St. Louis, on the 10th day of April, 1818, asks an outfit of one hundred brave companions, well equipped, to set out from Siberia in autumn, with rein-deer and sleighs, to pass over the ice of the frozen sea. On being furnished with an outfit of this description, he engages to explore the concave regions, and discover a warm, or at least a temperate country, of fertile soil, well stocked with animals and vegetables, if not men, on reaching about sixty-nine miles beyond latitude eighty-two degrees. The route, intended to be pursued by Captain Symmes, appears to be that of the rein-deer, and the time of setting out, the same season of the year, in which (according to Professor Adams) the rein-deer migrate from that coast north. In this route it would be necessary to cross the verge, or region of most intense cold, with the greatest possible expedition, so as to reach an inner temperate climate, in the shortest time. The concave regions could be partially explored during the winter; and the party return in the spring, and at the same time of the rein-deer, to the mouth of the river Lena.

The Russians have been making considerable exertions to explore the northern regions. Baron Wrangle made an attempt of this kind, in the year 1821. And a second attempt was made in the year 1822, by travelling with sledges, drawn by dogs.[47] But, probably owing to the party not having faith in the winding meridians about the verge of the polar opening, or being unacquainted with their direction according to the theory of concentric spheres, they were bewildered, and kept travelling in the neighbourhood of the verge, the region of greatest cold, instead of proceeding in a direct course towards the pole, until they were finally obliged to return without accomplishing the object of the expedition.

At the present time (August, 1824) an expedition is fitting out in Russia at great expense, under the auspices of that distinguished patron of science, Count Romanzoff, for the purpose of making discoveries in the northern regions, with the intention of exploring over land, or on the ice, as far as it may be found practicable. The celebrated Admiral Kruzenstern, is to exercise a general superintendance over the expedition, while the immediate command is to be conferred on some distinguished Russian officer.

The continent of North America, would, in my opinion, be a more suitable place, for an exploring party to set out from, than the coast of Siberia. A company of men, well armed, could travel over land, and draw their provisions and baggage on hand sledges, on the snow or ice, as Hearne did during his journey, with light canoes for the purpose of crossing rivers and lakes, should such be found to obstruct their progress. In this manner, the party would soon cross the verge, or "barren grounds," as Hearne calls it, and arrive in that country of abundant game, of which the Indians informed him. Hearne, according to his journal, reached nearly the seventy-second degree of north latitude, and his general course is laid down as being north-westwardly, from Fort Churchill to the mouth of Copper-Mine river, which he says disembogues itself into the Northern sea, flowing in a northerly direction. Me-lo-no-bee, the Indian chief, who served as Hearne's guide from Hudson's Bay, pointed out the mouth of Copper-Mine river, as being in a north-eastwardly direction from Fort Churchill, and flowing in an eastwardly course. Subsequent discoveries have, I believe, determined Me-lo-no-bee to be correct in this particular, as that river has been ascertained to empty into the waters of the Atlantic north of Repulse Bay, several hundred miles distant from where Hearne lays it down on his map. It is so laid down in the map accompanying Ross' voyage of discovery. How Hearne could be so much mistaken in the course he travelled, as to lay it down at nearly a right angle from its true course, is rather unaccountable: he must have been deceived by the winding meridians of the verge, which turned him to the right; when to have passed directly into the concave, he ought, on arriving at a certain point, to have proceeded west of north, then west, and finally south-west, which would probably have conducted him to that country, which the Indian represented as being far to the west, or south-west, and so warm that there was never any frost. In this direction, an exploring party ought most probably to travel, first north until they come to the verge; where (if they are on the continent of America) the meridians begin to wind to the right, then gradually, as they advanced, incline to the west, then true west, then south of west, and finally, when entirely beyond the apparent verge, to the south-west, if not due south. In crossing the verge, the cold would no doubt be considerable: but cold in those regions, as measured by the thermometer, appears to us much greater than the feelings of those exposed to that temperature indicate. Hence it was, no doubt, that Parry's crew could hunt in winter, when the medium was below zero. And the Russians set out on their expedition over the ice in 1821, when the cold was thirty-two degrees Reaumur; and this too accounts for Hearne's sleeping in the snow, without fire, by only digging a hole, and lying therein, with his sledge turned up to windward. It does not appear that he complained of excessive cold; though he travelled nearly all winter. He had also several Indian women in company. The regions through which he passed, as well as that in which Ross and Parry were, are alleged to be the coldest of the earth; and that those men experienced as great a degree of cold as would be in passing the verge into the concave regions.

But I am of opinion that the most practicable, the most expeditious, and the best mode of exploring the interior regions would be by sea, and by way of the south polar opening, crossing the verge at the low side, in the Indian ocean, where it is presumed the sea is always open, and nearly free from ice. But, as we are residents of the northern hemisphere, the nearness of the north polar opening to us, and the more immediate advantages which would result to us from an intercourse with the countries within the concave to the north, would seem to point out that as the most proper direction to be pursued; though the difficulties to be encountered in passing the verge of the north polar opening, would doubtless be much greater than those of the south, the cold much severer, and the ice more compact and difficult to pass. However, notwithstanding all these difficulties, the object, I think, might be safely accomplished by sailing, either east of Spitzbergen, or between Spitzbergen and Greenland; where, writers, in whom confidence may be placed, inform us, that the sea is open all winter. The greatest difficulty to be apprehended, would be the accumulation of drifting ice in the summer season; but in the winter, that difficulty, perhaps, would not be presented as in the fall or commencement of winter, the ice would attach itself to one shore or the other, and become permanent.

The Russians who wintered on Spitzbergen, say that the sea was open during the whole winter, quite across the north end of the island. Several sailors who were once left on an island near Spitsbergen, lived there several years; though destitute of almost every necessary of life, they were not only able to support the cold of the winters, but even to supply themselves with provisions, and light, in those dreary regions. They finally returned in health and safety to their native country and friends. This island is probably as cold as any spot that is known to our sphere.

A vessel, almost at any time in summer, could sail to, and remain at Spitzbergen, (having the necessary conveniences on board to make the crew comfortable) for two or three years. They could lie all winter at the north part of the island, and after being there long enough to become acquainted with the nature and changes in the sea to the north of them, they could take some favorable opportunity, and reach the pole, (if the earth be a globe) or the interior concave regions. The distance from the north of Spitzbergen to the pole is only six hundred geographical miles.

Another favorable direction for making the discovery is, by Bhering's straits on the north-west coast of America: And an additional advantage which is presented by this direction, is, that if the vessels should be obstructed by, or frozen in the ice, the party could proceed by land on the shore of America, (which is supposed to communicate with the concave regions,) a party remaining with the vessels till the others returned.

In case an expedition of discovery should be fitted out for the purpose of making the attempt, by either route, the safety of the party would require that two vessels should be equipped with rather more than an ordinary number of men, and with a double number of boats at least; some so light and portable as to be easily carried by men over ice, or necks of land, should it become necessary.

Vessels propelled by steam would be preferable to any other, as they could more easily avoid the floating ice in passing the verge; as, also ascend rapid rivers in the interior, should such be discovered, and it be found necessary to ascend them. The vessels should be equipped with masts, sails, and every part of rigging necessary for sailing; with a ballast of coal, which should not be used, or any other fuel for steam purposes, until they come within the neighbourhood of the ice, through which, by pursuing a proper course, it is believed, they would in a few days pass, and arrive at a more temperate climate, and a country where they would be abundantly supplied with both wood and provisions. Perhaps it would be advisable to take on board a small boat, with a proportionate steam-engine, for the purpose of running up shallow rivers, or along coasts, to make more minute observations.

But the most important matter of all to be observed, and that on which the success of the expedition must depend, would be a proper observance of the principles of the theory, and a due attention to the winding meridians, and curvatures of the parallels of latitude, when the verge shall be crossed; and which will require the party to be continually varying their course as they proceed forward in accordance with the place at which the attempt shall be made.

The expense of an expedition of this kind, would not be very great; at least not considerable when compared with the magnitude of the object to be accomplished, though I have not made, nor do I consider myself adequate to make minute estimates on the subject. But I should conclude that a sum of one or two hundred thousand dollars would be amply sufficient to defray all expenses attending such an expedition. Should an attempt be made by way of the south polar opening, with vessels fitted out as for a whaling voyage, the expense would probably not be the one fifth part of that sum. And were an expedition undertaken over land, from some post high north on the continent of America, the expense must be still less.

# CHAPTER X.

*A short Biographical sketch of Captain Symmes; with some observations on the treatment which he has met with in the advancement of his Theory.*

John Cleves Symmes, the author of the Theory of Concentric Spheres, is the son of Timothy Symmes, of the state of New-Jersey, whose father's name was also Timothy, and who was the son of the Rev. Thomas Symmes, of Bradford, who graduated at Harvard college, in 1698. Mr. Elliot, publisher of the New-England Biographical Dictionary, at Boston, in the year 1809, makes honourable mention of his name. Timothy Symmes, the grandfather of the subject of this sketch, had but two sons; the one, John Cleves Symmes, well known as the father and founder of the first settlements in the Miami country; and the other, Timothy, the father of our Theorist, and from whom the present family of Symmes, in the Miami country, are descended.

Captain Symmes is now about forty-six years of age. He is of middle stature, and tolerably proportioned; with scarcely any thing in his exterior to characterize the secret operations of his mind, except an abstraction, which, from attentive inspection, is found seated on a slightly contracted brow; and the glances of a bright blue eye, that often seems fixed on something beyond immediate surrounding objects. His head is round, and his face rather small and oval. His voice is somewhat nasal, and he speaks hesitatingly and with apparent labour. His manners are plain, and remarkable for native simplicity. He is a native of the state of New-Jersey. During the early part of his life, he received, what was then considered, a common English education, which in after life he improved by having access to tolerably well selected libraries; and being endued, by nature, with an insatiable desire for knowledge of all kinds, he thus had, during the greater part of his life, ample opportunities to indulge it.

In the year 1802, and at the age of about twenty-two years, Mr. Symmes entered the army of the United States, in the office of ensign; from which he afterwards rose to that of captain. He continued in service until after the close of the late war with Great-Britain. While attached to the army he was universally esteemed a brave soldier, and a zealous and faithful officer. He was in the memorable battle of Bridgewater; and was senior Captain in the regiment to which he belonged. The company under his immediate command, that day, discharged seventy rounds of cartridges, and repelled three desperate charges of the bayonet.

Afterwards, in the sortie from Fort Erie, Captain Symmes, with his command, captured the enemy's battery number two; and with his own hand spiked the cannon it contained: yet, owing to the want of correct information, or from some other cause, the honour and the reward of this achievement, were alike bestowed upon others. And, it is a fact not less to be regretted, that the official report of the battle of Bridgewater, has represented the regiment, to which Captain Symmes was attached, as almost the only one that retreated at Lunday's lane; when, in truth, it was nearly the only one which uniformly maintained the positions it was *ordered* to maintain, throughout the action. Captain Symmes, has since, however, substantiated the correctness of its conduct, by obtaining the necessary acknowledgments; some of the particulars of which were communicated to the Historical Society of New-York, and published, in the newspapers of the day. The truth of this statement, has also been confirmed to me, by a respectable Officer, who was in the action, and witnessed the occurrence.

During the period of about three years, immediately after the close of the war, and after Captain Symmes had left the army, he was engaged in the difficult and laborious task of furnishing supplies to the troops stationed on the upper Mississippi. How he succeeded in this business I am not informed; but, I conclude from his present circumstances, that he could not have realized any very considerable pecuniary advantage from the enterprise. Since that time he has resided at Newport, Kentucky; devoting, almost exclusively, the whole of his time and attention to the investigation and perfection of his favourite Theory of Concentric Spheres.

In a short circular, dated at St. Louis, in 1818, Captain Symmes first promulgated the fundamental principles of his theory to the world. He addressed a copy to every learned institution, and to every considerable town and village, as well as distinguished individuals, of which he could gain any intelligence, throughout the United States, and to several learned societies in Europe.

The reception this circular met with, was that of ridicule; it being looked upon as the production of a distempered imagination, or the ravings of partial insanity. Indeed, it became a fruitful source of jest and levity, to publishers of the public prints of the day generally, all over the Union. The Academy of Sciences in Paris, before which it was laid by Count Volney, decided that it was unworthy of their consideration; and the editor of the London Morning Chronicle, could not be induced to credit the statements of respectable men, who declared that Symmes was not a madman. But in this, his fate is not peculiar. The experience of the world has taught us, that the authors of new doctrines, have mostly shared a similar lot. An excellent contemporary writer has remarked, that, "the fate of many projectors have been so melancholy, that it requires, at this day, the daring spirit, and the enthusiasm which are naturally allied to genius, in any man to announce himself as the inventor of any thing new and extraordinary. The patience and perseverance of a Galileo, and the adventurous spirit of a Fulton, are necessary to him who would benefit his species by the results of original plans and forms, or that of new combinations of old and tried ones. Hence we cannot but respect and admire the man, who, regardless of the hard fate of so many who have trod before him, in the thorny path of improvement, still has the fortitude and philosophy of mind to spend years in toil and study—to labour by day with persevering industry—and trim the midnight lamp with the vigilance ascribed to the ancient vestals, in bringing to perfection an idea, from which he hopes to reap fame and benefit to himself, and to reflect credit, at the same time, on the genius of his country."

Captain Symmes published two other numbers at St. Louis, in the year 1818; the one went to prove, by geometrical principles, that matter must necessarily form itself into concentric spheres, and the other treated of geological principles. His two next numbers, marked four and five, (the one treating of the original formation of the Allegheny mountains, and the other claiming the discovery of open poles,) I have never had an opportunity of seeing. His sixth number appeared, dated at Cincinnati, in January, 1819, which contains a number of items and principles that he proposes treating of in subsequent numbers. His seventh number, entitled "*Arctic Memoir*," is dated at Cincinnati, in February, 1819; and another number, entitled "*Light between the Spheres*," dated at Cincinnati, in August, 1819, was published in the National Intelligencer. From that time to the present, numerous pieces from the pen of Captain Symmes have appeared in different newspapers; but the most prominent and grand doctrines, on which his theory is based, are contained in the papers above enumerated. Independent of his written publications, he has delivered a number of lectures on the theory,—first at Cincinnati, in 1820, and afterwards at Lexington and Frankfort, in Kentucky, and at Hamilton and Zanesville, in the state of Ohio. Several of these lectures I had the pleasure of hearing; and the respectable number of auditors, and the profound stillness that reigned, evinced in the strongest manner the interest felt by all present in the subject. In addition to the various facts and phenomena, to which he adverts in support of his positions, he delineates in his lectures, upon a wooden sphere, constructed on the principles of his theory, the cause of the winding meridians, the icy hoop or verge, and the course which ought to be pursued to reach the interior regions, with the confidence of mathematical certainty.

Captain Symmes's want of a classical education, and philosophic attainments, perhaps, unfits him for the office of a lecturer. But, his arguments being presented in confused array, and clothed in homely phraseology, can furnish no objection to the soundness of his doctrines. The imperfection of his style, and the inelegance of his manner, may be deplored; but, certainly, constitute no proof of the inadequacy of his reasoning, or the absurdity of his deductions. There is scarcely a single individual, with whom I have conversed, who does not confess that, if the facts which he adduces, and the arguments he uses, were handled by an able orator, they would produce a powerful effect. In short, those who attend to his lectures, without regarding his peculiarities of style and manner; who reflect alone on their substantial parts, without regarding the want of eloquence in the lecturer; who presume to think for themselves, and are able to comprehend the naked facts, and unadorned arguments, which he advances, will not fail to discover in them many particulars well worthy of their consideration; and many arguments calculated to stagger their faith in pre-conceived opinions.

In the year 1822, Captain Symmes petitioned the Congress of the United States, setting forth, in the first place, his belief of the existence of a habitable and accessible concave to this globe; his desire to embark on a voyage of discovery to one or other of the polar regions; his belief in the great profit and honour his country would derive from such discovery;—and prayed that Congress would equip and fit out for the expedition, two vessels of two hundred and fifty, or three hundred, tons burthen; and grant such other aid as government might deem necessary to promote the object. This petition was presented in the Senate by Col. Richard M. Johnston, a member from Kentucky, on the 7th day of March, 1822; when, (a motion to refer it to the committee of Foreign Relations having failed,) after a few remarks it was laid on the table.—*Ayes*, 25.

In December, 1823, he forwarded similar petitions to both houses of Congress, which met with a similar fate.

In January, 1824, he petitioned the General Assembly of the state of Ohio, praying that body to pass a resolution approbatory of his theory; and to recommend him to Congress for an outfit suitable to the enterprise. This memorial was presented by Micajah T. Williams; and, on motion, the further consideration thereof was indefinitely postponed.[48]

That Captain Symmes is a highminded, honorable man, is attested by all who know him. He has devised a theory whereby to account for various singular and interesting phenomena; and more satisfactorily to explain a great variety of acknowledged facts.

He argues from the effect to the cause, in many of his positions, with great perspicuity. And the circumstance that few of the learned have yet attempted to show that his principles are founded in absurdity, should at least entitle him to the respect, and his theory to the attention, of every candid man. Notwithstanding he has been buffeted by the ridicule and sarcasm of an opposing world for seven years, under great pecuniary embarrassments; he still labours with unshaken faith, and unbroken perseverance; with a willingness at any time to test the truth of his speculations amid the icy mountains of the polar seas.

Already has he passed the meridian of life; and should he be called from time, without establishing his theory by actual discovery; the science he has embodied, and the facts he has collected and arranged in support of it, together with his undeviating and indefatigable industry, in the face of

"The world's dread laugh, which scarce

The firm philosopher can scorn,"

will bear a testimonial to his talents and worth, that the best of his species will ever delight to acknowledge. And though he may not have accounted for every particular, or brought forward every argument that might possibly be advanced in support of his positions; he has, nevertheless, collected a greater number of peculiarly interesting facts, and embodied a stronger phalanx of proof, than could well have been expected on a subject so new, and in the hands of the original discoverer.

If, hereafter, it should be ascertained that Symmes's Theory of the Earth is true, impartial posterity will not withhold the honour and fame due to the name of the discoverer.

It is hoped, however, that the present age will not so far forfeit to posterity the high character it now sustains in scientific discovery, as to remain deaf to his solicitations; but, that the citizens of our own country in particular, if not the whole world, will unite in testing the truth of his principles; and in doing justice to the merits of this extraordinary man.

**FINIS.**

## FOOTNOTES

[1] National Intelligencer of June 10th, 1824.

[2] A tolerably correct representation of the sphere might be made by taking a hollow terrestrial globe, such as are used in colleges, and insert a saw at north latitude sixty-eight degrees in Lapland, sawing obliquely

through, so as to come out at latitude fifty degrees in the Pacific ocean. The aperture thus produced, will show the general dimensions and slope of the north polar opening. And in the southern hemisphere, commencing with the saw at south latitude thirty four degrees, in longitude between fifty and fifty-five degrees east, in the Indian ocean, and sawing obliquely through, in the same manner, so as to come out at south latitude forty-six degrees, and longitude one hundred and thirty degrees west, in the South Pacific ocean, you will represent the appearance of the south polar opening; and the whole will exhibit a general representation of the sphere, according to the new theory.

[3] The solidity of the earth is easily calculated by the measure of a meridional degree; but the result will be different according to the measurement assumed, as the length of a degree differs in different latitudes. "Notwithstanding all the admeasurements that have hitherto been made, it has never been demonstrated, in a satisfactory manner, that the earth is strictly a spheroid; indeed, from observations made in different parts of the earth, it appears that its figure is by no means that of a regular spheroid, nor that of any other known regular mathematical figure; and the only certain conclusions that can be drawn from the works of the several gentlemen employed to measure the earth is, that the earth is something more flat at the poles than at the equator." [Keith on globes p. 56. New-York, 1811.]

According to Mott's translation of Newton's Principia, book 3, page 243, the equatorial diameter of the earth is 7964 English miles, and the polar diameter 7929, for as 230:229::7964:7929 miles, the polar axis.

Cassini, who adopted Picard's measure of a degree, makes the diameter of the earth 7967 statute miles; others have estimated it at 7917, and some at 7910 miles. But the estimate which is now esteemed most correct, I believe, is, that the equatorial diameter is 7977 English miles, and the polar diameter 7940. From this we may ascertain the solid contents of the earth. The axis of the earth then assumed to be 7940 and 7977 miles respectively, the area of the generating eclipse is (7940 × 7977 × 0,7854=) 49745178,252: and its area multiplied by two thirds of the longer axis, gives the solidity equal to (49745178,252 × 2/3 × 7977=) 264544857944,136 cubic miles.

[4] Physical World, p. 42.—Adam's Philosophy, vol. 4, p. 206; Philadelphia, 1807.

[5] "Dr. Herschel has observed a faint illumination in the unlighted part of the planet Venus, which he ascribes to some phosphoric quality of its atmosphere." Editor's note to Adams' Philosophy, vol. 4, p. 204, Philadelphia, 1807.

*Quere*—Might not such an appearance be accounted for as rationally, by supposing the rays of the sun to shine or be reflected, through one of her polar openings, and fall on the verge of the sphere at the opposite polar opening?

[6] Hearne's Journal, pp. 357, 368.

[7] Hearne's Journal, pp. 364, 365.

[8] Hearne's Journal, p. 442.

[9] Hearne's Journal, p. 442, 443, 444, 445, 446.

[10] Ibid, p. 445.

[11] Barrington and Beaufoy, pp. 21, 51.

[12] Ibid, pp. 25, 61.

[13] Ibid, pp 25, 32, 37, 61.

[14] *From the National Intelligencer of Sept. 30, 1824.*

"POLAR SEAS.—The fact that there are open seas round both the earth's poles, has received strong corroboration within the last few months. We have now a letter on our table from a naval officer at Drontheim, who notices the fact that Captain Sabine had good weather, and reached eighty degrees and thirty-one minutes north latitude, without obstruction from the ice; so that the expedition might easily have proceeded farther had its object so required. We have also had the pleasure to meet recently with a British officer who, with two vessels under his command, last season penetrated to seventy-four degrees twenty-five minutes south latitude, in the antarctic circle, which is about three degrees beyond Cook's utmost limit. There he found the sea perfectly clear of ice, and might have prosecuted his voyage towards the pole, if other considerations had permitted. There was no field ice in sight towards the south; and the water was inhabited by many finned and hump-backed whales; the longitude was between the south Shetland Islands, lately discovered, and Sandwich land: this proves the former to be an Archipelago (as was supposed) and not a continent. The voyage is remarkable as being the utmost south upon record; and we hope to be favoured with other particulars of it. At present we have only to add, that the variation of the needle was extraordinary, and the more important as they could not readily be explained by the philosophical principles at present maintained on the subject."

*Literary Gazette.*

[15] Barrington and Beaufoy, p. 74.

[16] Purchas, vol. 1, p. 479.

[17] Hearne's Journal, p. 7.

[18] Barrington and Beaufoy, p. 36.—Dr. Birch's history of the Royal Society, vol. et seq.

[19] Bar. p. 101.

[20] Barrington and Beaufoy, pp. 25, 124.

[21] At the mouth of St. Peter's river, in winter, it is as much colder than at Sacket's Harbour, as Sacket's Harbour is colder than Mobile, although St. Peter's is west and Mobile south of Sacket's Harbour, at nearly equal distances.

[22] Barrington and Beaufoy, p. 74.—Ross' Voyage, vol. 1, p. 52, London, 1819.

[23] Darwin's Botanic Garden.

[24] Ross' Voyage, v. 1, p. 175.

[25] Ross' Voyage, v. 1, p. 110.

[26] Ross' Voyages, v. 1, pp. 71, 135, 199, 206.

[27] Mackenzie states, "that sometimes the land *looms*, so that there may be a great deception in the distances."—Mackenzie's Voyage, p. 11, New-York, 1802.

[28] Ross' Voyage, v. 2, p. 4, London, 1819.

[29] Ross' Voyage, v. 2, p. 119.

[30] Physical World, p. 72.

[31] EARTHQUAKES.—M. Biot, after detailing the phenomena of the earthquake, on the 22d of February, 1822, concludes an interesting paper with these observations:—

In the infancy of Chemistry and Natural Philosophy, it was imagined that earthquakes might be easily explained; in proportion as these sciences have become more correct and more profound, this confidence has decreased. But by a propensity, for which the character of the human mind sufficiently accounts, all the new physical agents which have been successively discovered, such as electricity, magnetism, the inflammation of gases, the decomposition and recomposition of water, have been maintained in theories as the causes of the great phenomena of nature. Now all these conjectures seem to be insufficient to explain convulsions so extensive, produced at the same time over such large portions of the earth, as those which take place during earthquakes. The most probable opinion, the only one which seems to us to reconcile, in a certain degree, the energy, the extent of these phenomena, and often their frightful correspondence in the most distant countries of the globe, would be to suppose, conformably to many other physical indications, that the solid surface on which we live is but of inconsiderable thickness in comparison with the semi-diameter of the terrestrial globe; is in some measure only a recent shell, covering a liquid nucleus, perhaps still in a state of ignition, in which great chemical or physical phenomena operating at intervals cause those agitations which are transmitted to us. The countries where the superficial crust is less thick or less strong, or more recently or more imperfectly consolidated, would, agreeably to this hypothesis, be those the most liable to be convulsed and broken by the violence of these internal explosions. Now if we compare together the experiments on the length of the pendulum, which have been made for some years past with great accuracy, from the north of Scotland to the south of Spain, we readily perceive that the intensity of gravitation decreases on this space, as we go from the Pole towards the Equator, more rapidly than it ought to do upon an ellipsoid, the concentric and similar strata of which should have equal densities at equal depths; and the deviation is especially sensible about the middle of France, where too there has been observed a striking irregularity in the length of the degrees of the earth. This local decrease of gravity in these countries should seem to indicate, with some probability, that the strata near the surface must be less dense there than elsewhere, and perhaps have in their interior immense cavities. This would account for the existence of the numerous volcanos of which these strata show the traces, and explain why they are even now, at intervals, the focus of subterraneous convulsions.

[32] Cook's Geography, v. 2, p. 250—Also Rees' Cyclopedia, article Lake.

[33] Rees' Cyclopedia, article Lake Geneva.

[34] Rees' Cyclopedia, article Lake.

[35] Ross' Voyage, v. 1, p. 225.

[36] Ibid, v. 1, p. 144.

[37] Humboldt.

[38] Physical World, p. 105.

[39] Barrington and Beaufoy, p. 106, and Purchas, v. 3, pp. 499, 500.

[40] The late George Adams, in his Philosophy, treating of refraction, states, that "at the horizon, in this climate, (England) it is found to be about thirty-three minutes. In climates near the equator, where the air is pure, the refraction is less; and in the colder climates, nearer the pole, it increases exceedingly, and is a happy provision for lengthening the appearance of the light at those regions so remote from the sun. Gassendees relates, that

some Hollanders, who wintered in Nova-Zembla, in latitude seventy-five degrees, were agreeably surprised with a sight of the sun seventeen days before they expected him in the horizon. This difference was owing to the refraction of the atmosphere in that latitude."—Adams' Philosophy, v. 4, p. 112, Philadelphia, 1807.

[41] Keith on Globes, p. 144.—

[42] Ibid, p. 149.

[43] Sir Isaac Newton, in his Principia, under prop. 16, book 3, lays down the following proposition, viz: that "*the heat of the sun is as the density of his rays, that is reciprocally as the squares of the distances from the sun.*" From this principle, it has been assumed by some of our modern astronomers, that but few of the planets can be inhabited, as if the effect of light and heat are reciprocally proportionate to the squares of the distances from the centre of their propagation; and if you divide the square of the earth's distance from the sun, the quotient will show, that the light and heat, which Mercury receives, are about seven times greater, making it more than twice as hot as boiling water. The light and heat communicated to Saturn, being only the one hundredth part of that of the earth, the difference is more than seven times as great as that between our summer heat and red hot iron, if the light and heat of the sun are only in proportion to the density of his rays. Such extremes of heat and cold, we would naturally conclude must totally preclude all material being, if in the least degree resembling those we are acquainted with; nor could any of the vegetable world, known to us, germinate in either extreme; nay, even the matter of our globe would scarcely withstand it, our oceans would be dissipated in vapour, on Mercury, and frozen to the bottom on Saturn. Considerations like these must induce us to conclude, that light and heat cannot be communicated exactly on the plan laid down by Newton, viz: that the heat of the sun is simply as the density of his rays: for though the sun's rays may be the *sine qua non*, without which no light or heat would be communicated, yet the *quantum* of heat may depend on the density and co-operation of the medium through which it passes, or upon some other circumstance not known to us, and perhaps impossible for us to know.

[44] I am indebted to an excellent Hebrew scholar for the following:

NOTE. The words תהו ובהו *Theoo* and *Beoo*, (Genesis, chapter 1, verse 2,) which has been rendered by the translators of our bible, "Without form and void," might perhaps, with equal propriety, have been translated "without form and hollow."

1. *Theoo*, the root, agreeably to the Hebrew grammar, is found as a noun תה־ or תהה *The* or *Thee*, and, is rendered *confusion*, loose, unconnected, without form, order, or the like; and so well understood.

2. *Be-oo*, the root, is, according to the same rule, found in בה—*Be*, (*Bethhey*) *hollow;* it occurs not only in this form but—

1. As a noun בהו Beoo—Hollow, empty, having nothing in it but air, filled only *vacuo aere*, with empty air, as Lucan calls it, Lib. 5, line 94.

2. As a noun fem: in reg: בת, בת עין Bet, Bethoin, the apparent hollow, or pupil of the eye, &c. Comp. בבת Bebath, under, בב Beb.

3. As a noun fem: תבה *Thebe* in Reg: תבת Thebeth, an ark, a hollow vessel, under 2d head of בב Beb. occurs not as a verb in kab, but

1. As a *participial* noun, or participle in Nipth נבוב Neboob, hollow, made hollow, &c.

2. It is applied spiritually, hollow, empty, vain.

3. To the sight, or pupil of the eye; that part of the eye which appears hollow, and admits the light. See Parkhurst's Hebrew Lexicon.

Had the learned translators of our bible possessed a knowledge of the theory of concentric spheres, it is probable they would have given the English reader the most correct meaning of the words, תהו ובהו "*without form and hollow*," or "*shapeless and hollow*."

[45] The application which the Dr. makes of this structure of the earth is this: that the concave sides of the spheres are made up of magnetic matter; that they revolve about their diurnal axes in about twenty-four hours; that the outer sphere moves either a little faster or a little slower than the internal ball; that the magnetic pole, both of the external shell and included globe, are distant from the poles of rotation; and that the variation arises from a change of the relative distances of the external and internal poles in consequence of the difference of their revolutions. [See life of Dr. Halley.]

In Rees' Cyclopedia, under the article 'ring,' is the following sentence; by which it appears that Kepler first suggested the earth to be composed of concentric crusts. "Kepler, in his Epitom. Astron. Copern. (as after him Dr. Halley, in his enquiry into the causes of the variation of the needle, Phil. Trans. No. 195.) supposes our earth may be composed of several *crusts* or *shells*, one within another, and concentric to each other. If this be the case, it is possible the ring of Saturn may be the fragment or remaining ruin of his former exterior shell, the rest of which is broken or fallen down upon the body of the planet."

[46] Maclaurin, in his fourteenth chapter of the second volume on Fluxions, investigates the theory of Dr. Halley at considerable length; and in conclusion, appears to consider the existence of a hollow globe as very possible.

[47] *From a London paper, under the head of*

"RUSSIAN DISCOVERIES.—In the year 1820, a journey of discovery, by land, was ordered by the government, to explore the extreme north and north-east of Asia.—Lieutenants Wrangle and Anjou, of the navy, were chosen for this expedition. After having made the necessary preparations, they departed from Neukolyma, in the north-eastern part of Siberia, on the 19th of Feb. 1821, in sledges drawn by dogs, when the cold was thirty-two degrees Reaumur, in order to ascertain the position of Schehaladshoi-Noss, which captain Burney conjectured might be an isthmus, joining Asia with the continent of America. The travellers succeeded in determining the whole coast astronomically, going themselves entirely round the coast, and proceeding a day's journey farther to the west; thus convincing themselves that Asia and America are not united there by an isthmus. On the 13th of March, the expedition returned to Neukolyma. On the 22d of March, Mr. Wrangle undertook another journey, likewise on sledges drawn by dogs, with ten companions, in the direction to the North Pole, in order to look for the great continent which is supposed to exist there. The principal obstacle they met with, was thin ice, which being broken to pieces by continued storms, was piled up in mountains, and rendered farther progress impossible. At a bear hunt, which the company undertook, they observed a sudden bursting of the ice, accompanied with a dreadful noise resembling thunder. On their journey back, which the travellers were obliged to make without accomplishing their object, they surveyed the bear islands, and after an absence of thirty-eight days, arrived safely at Neukolyma on the 28th April, where they are to remain for the year 1822, and then to continue their researches."

[48] Journal of the House of Representatives of Ohio; session of 1823, '24—p. 224.

www.ingramcontent.com/pod-product-compliance
Ingram Content Group UK Ltd.
Pitfield, Milton Keynes, MK11 3LW, UK
UKHW031338260325
456749UK00002B/341